T0315275

Matter

Matter

The Magnificent Illusion

GUIDO TONELLI

Translated by Edward Williams

polity

First published in Italian as *Materia. La magnifica illusione* (Varia collection).

The translation of this work has been funded by SEPS
Segretariato Europeo per le Pubblicazioni Scientifiche

Via Val d'Aposa 7 - 40123 Bologna - Italy
seps@seps.it - www.seps.it

The translator would like to thank Dr Ryan Buckingham for his invaluable help in preparing the manuscript.

Polity Press
65 Bridge Street
Cambridge CB2 1UR, UK

Polity Press
111 River Street
Hoboken, NJ 07030, USA

ISBN-13: 978-1-5095-6414-9 – hardback

A catalogue record for this book is available from the British Library.

Library of Congress Control Number: 2024934246

Typeset in 11 on 14pt Warnock Pro
by Cheshire Typesetting Ltd, Cuddington, Cheshire
Printed and bound in Great Britain by CPI Group (UK) Ltd, Croydon

For further information on Polity, visit our website:
politybooks.com

To the sweetest little Leon

"Matter is the great illusion. That is, matter manifests itself in form, and form is apparitional."

JACK LONDON

"We are such stuff as dreams are made on and our little life is rounded with a sleep."

WILLIAM SHAKESPEARE

Contents

Acknowledgements

I wish to thank my dear friend, the great historian, Antonello Mattone, who helped me to relive Poggio Bracciolini's cinematographic life, by telling me his story as a direct witness to the events.

A special thought for Emanuele Montibeller, who offered me the gift of putting me in touch with Michelangelo Pistoletto. I am eternally grateful to him for this and for our many chats about art and the life we have spent together.

I consider it an immense privilege having been able to spend time in recent years with maestro Pistoletto and the wonderful Maria. The long conversations we have had provided the inspiration for many parts of this book.

A particular thank you to Quirino Principe who introduced me to aspects of the world of music whose existence I hadn't even suspected.

And finally immense gratitude to Luciana, my precious life partner, for having read and critiqued the whole manuscript, endlessly spurring me on to make it ever clearer and more incisive. Without her, not only would this book never have seen the light of day, but my own life would have been distinctly greyer and less meaningful.

Prologue

Posara, Tuscany, 11 August 1945

He'd finished the uphill stretch now. All that was left was the road down from Moncigoli to their village. He was looking forward to getting there and telling everyone the great news. They'd done it, he and his brother-in-law Attilio, the best mechanic in town. They'd found a house and signed the rental agreement.

The flat was in the centre of La Spezia, in a building on the corner of Corso Cavour and Via di Monale. It was beautiful and sufficiently spacious to accommodate their two families: nine people in all, plus a baby on the way, given that Anita, his wife, was pregnant. They had to sort out the sleeping arrangements for the three bedrooms, since the main room would serve as a tailor's workshop. With the end of the war, people were returning to work and maybe, if things went well, they would soon need to employ a dressmaker or two.

The ride downhill was easy; the rod brakes of his sturdy Atala, the bicycle which had been his faithful companion through these difficult years, functioned perfectly in slowing down the movement into each bend. He'd set out that morning from

Posara, the tiny village near Fivizzano where they'd moved to survive the war and had effortlessly covered the forty or so kilometres from there to La Spezia. It was a journey he'd done many times. He knew every bend by heart.

That black bicycle, with its crankcase protecting the chain, its bell with the die-cast maker's mark and the dynamo which allowed him to cycle even when it was getting dark, had been an essential resource in ensuring his family's survival. In the surrounding villages there was always one peasant farmer or another needing to turn a worn-out greatcoat inside out or mend the holes in a jacket which he had to wear for a wedding. He would hurry off and return home with eggs or a bottle of fresh milk. Everyone knew the tailor who travelled from village to village on his bicycle.

His wanderings soon became the ideal cover for taking messages to the partisan groups that were operating in the area. A note would be handed to him in the evening and all he had to do was remove the bike saddle and slip it into the seat post. On several occasions he had even read the notes, but the messages were incomprehensible; sometimes they were phrases in code which indicated the arrival of columns of Nazis or fascists from Massa to carry out a round-up. Or they simply contained dates and numbers, effectively the coordinates for weapon launches or Allied aerial supply drops for the partisans.

The tailor had been lucky; nobody had betrayed him, and he had never been found out. On a couple of occasions, he had even been able to embrace his brother Giuseppe, a political commissioner of a partisan detachment of the "Apuania" Garibaldi Brigade. Giuseppe had given him a pistol, a Luger P08, supplied with calibre 9 Parabellum bullets, which he'd seized from a Wehrmacht soldier who'd fallen in an exchange of fire. The tailor did not like weapons, and he had rushed back to Posara in terror of being stopped and shot dead at the first Brigate Nere checkpoint. But everything had gone smoothly and, once he was back home, he had taken care to hide the pistol. He'd wrapped it in a filthy rag and buried it in the cowshed, right under the animals' feeding trough, where the hay was deepest, and he never touched it again.

He'd moved to Posara with his entire family at the beginning of 1942, as soon as it became clear that the tailoring business was definitively on hold. Nobody orders a new suit in wartime. In La Spezia everything that could be considered food had become impossible to find and too expensive, and with five mouths to feed he couldn't afford to take risks. And so, the family had evacuated to the countryside, to his recently deceased father's home village where his brothers and their families still lived. With their few furnishings of any value loaded onto a cart, they were all off to live in a single room set up above the cowshed. It was a kind of hayloft where they'd placed the solid wooden table from the tailor-shop, along with a portable sink for washing and three beds, where they would sleep in pairs around a stove which served for cooking and heating. For their physical needs, they went outside to a little wooden hut, where the sewage was collected to be recycled as fertilizer in the vegetable garden.

The peasant house was poor, the winters in the foothills of the Apennines were freezing, but there was no shortage of firewood in the woods of Turkey oak which surrounded the village. Higher up, in the autumn, they gathered chestnuts which, if dried, could be ground into flour. Everything else came from the animals: rabbits, chickens, three cows and a family of pigs. In the fields and kitchen gardens they grew potatoes, beans, cabbages, maize and various other vegetables. And there was plenty of fruit too, in season. A tough life for everyone, basically, but nobody starved.

The tailor often went to La Spezia, about once a month. He went to get general foodstuffs with the family's ration coupons or to exchange produce from the countryside for packs of flour or pasta. He also took the opportunity to take some food supplies to Giulia, his mother-in-law, a powerful, stubborn woman, who had refused to have anything to do with the evacuation to Posara with the rest of the family.

They'd tried everything to convince her, but without success. She'd stayed behind, living alone, in a sad, dark basement in Via Napoli, convinced that, old as she was now, nothing really bad could happen to her. She'd been a widow for many years and was used to living completely independently. You could still see, in her facial features, the now faded signs of a stunning beauty; she was always smiling, and she never left the house without perfect make-up. Occasionally she would return to that basement, before the curfew, in the company of an elderly suitor. Giulia preferred to fast rather than do without lipstick.

When her son-in-law arrived, it was always time for celebration, because the tailor had knocked her up a new blouse out of some threadbare sheets, or he'd bring her a little cheese and some fresh eggs and would give her news of Anita and Giuliano, of Marisa and the other children.

Giulia hadn't even been fazed when, on 19 April 1943, all hell had broken out over La Spezia. In the following days, using 173 Lancaster and five Halifax bombers, the RAF had dropped 1,300 tons of bombs, with the aim of destroying the naval base and the vast military arsenal, which was responsible for repairing the fleet. The bombs had instead devastated the historic centre and caused more than 120 deaths and almost a thousand injuries.

Even the building where Giulia was living was one of those bombed, but, miraculously, she was spared. The rescue teams had found her, with an elderly friend, in the basement, which she hadn't left to run to one of the shelters when the alarm had gone off. Many years later, she would confess that they hadn't actually heard the sirens, because her friend had brought with him one very last bottle of wine from his cellar which he thought had already been empty for some time and . . .

The tailor was thinking about all this as he went round the final bends. He was smiling, to himself, remembering Giulia's flamboyance. He was happy because they were going back to the city, to his city. Anita and the children would welcome him with joy. A new life was beginning for all of them.

The five years of the war had been dreadful. In the last phase, in particular, the whole area had been subject to the bloody acts of the Walter Reder SS division and the Brigate Nere from Massa. In the Summer of '44, first at Sant'Anna di Stazzema, then at Vinca and in scores of other small towns in the surrounding area, they had exterminated more than 800 old people, women and children. The situation had become too dangerous for Giuliano, the eldest son, who had crossed the front on Christmas night in '44 to join the Americans.

The tailor himself had been lucky to save his skin too, when he'd been stopped by the X Mas military groups. It'd happened on 10 January 1944, when he'd been cycling home from one of his trips to La Spezia and had run into a mobile checkpoint. The partisan action groups (GAP) operating in the city had attacked a tram transporting X Mas officers and men, causing deaths and dozens of injuries. The checkpoints set up at all the entrances to the city were used to round up men who were caught on the street. The tailor was arrested immediately, his bicycle was confiscated, and he was shoved into a large room, at the back of the bunker which served as the base for the checkpoint, together with another ten or so hopeless cases. The militiamen who were guarding them were very young and extremely nervous; they had their fingers on the machine-gun triggers and were shouting non-stop. The tailor was terrified. They could shoot them straightaway, or after they'd tortured them. In the best-case scenario, they would take them to jail and then transfer them to some prison camp in Germany. He was in despair at the idea that he might never see his family again.

After some hours of anxious waiting, a young officer came in, barked out his name and took him outside at gunpoint. The tailor was ready to be shot when he heard his name being called. "What, you don't recognize me? It's Antenore, Nives' son, your nephew."

The tailor's family were all communists and opponents of the regime. One of his brothers had emigrated to France to avoid arrest, because he had opposed the early actions of the Fasci di Combattimento. The rest of the family was made up of union members, artisans and workers, secretly members of the Communist Party and resistance fighters. With one exception: his cousin Nives' son, a hot-head, a good-looking young man, intelligent and playful whom he remembered well when, as a child, he'd come into the tailor's workshop with his mother, and he'd made him a little sailor-suit. Antenore had volunteered to join up with the Fascists as an act of rebellion, maybe out of a misguided sense of patriotism, and had been part of the Salò Republic. His family had wiped him from their memory. Nobody mentioned him again after they'd seen him marching past in uniform with Junio Valerio Borghese's militiamen, the troops responsible for the most ferocious reprisals against the partisans. Chance would have it that Antenore was part of that patrol. Even before the tailor realized what was happening, he had led him out of the bunker and, taking advantage of the darkness which had fallen, had put him back onto his bicycle and pushed him off down the slope sending him away with an affectionate: "Good luck, uncle."

The tailor has by now reached the last bend in the road and is feeling that all this is behind him. He can't wait to get back to his family; the first houses in the village are already in sight. Enough tragedies. Enough deaths. All he wants is to celebrate, run up the steps and take Anita in his arms and carry her around the room with the other children gathered around them . . .

The lorry heading up towards Moncigoli is carrying furniture. It's the only lorry driving around those roads. The driver is in a good mood and singing as he turns the wheel to tackle the first bend. He has no inkling that the cyclist, who appears out of nowhere heading towards him downhill, will in the blink of an eye end up under the wheels of his lorry and won't make it out.

The tragic end of the tailor, dead before his forty-fifth birthday, marked the whole family, forever. The news of the accident had spread immediately through the little valley. His eldest son arrived at the scene of the accident, having run as if possessed to that cursed bend, a few hundred metres from the village. Yet there was nothing he could do but wrap his arms around his father's body, which had been lying there, lifeless, on the tarmac, his face disfigured, and howl in despair.

All the dreams, the hopes of a better life shattered instantly. The lad, not yet twenty, won't have time to mourn. It will fall to him to take on the burden of the family, his mother and four younger siblings, including the one due to be born in a few months' time.

Years will pass before a smile appears on his lips again. And only when, for the first time, he holds in his arms the son his wife Lea has just given birth to. Five years since the accident, Giuliano decides to name the little one after his father, the tailor: Guido Tonelli.

1

The mother of everything

The word matter contains the Latin word *mater*, mother, which seems to indicate its role as a primordial element at the origin of everything. In fact, its etymology conceals a number of nuances and is rich in multiple meanings.

When we think of matter, we are principally thinking of inorganic matter and thus imagining something inert and essentially arid. We fall into the trap of judging matter as something different from what we humans are, always rather presumptuous in considering ourselves to be made of a substance which has nothing to do with the ordinary. It almost seems as if the question of matter has little to do with us, as if we were made of a different substance, which is much nobler, so-called animate matter.

This is a centuries old assumption, which has given rise not only to great architectures of thought but also to arrogant attitudes which have made human beings prisoners of an infinite series of misunderstandings worth dwelling on. Everything stems from the role played, both in our lives and in our conception of the world, by our bodies, that is to say by the matter of which we are made, a role which we often tend to overlook, or even airbrush out completely.

A word with very deep roots

The Greek word corresponding to the Latin *materia* is ὕλη (*hyle*), among whose meanings we find wood, timber. It's the same etymological root as the Latin word *silva,* which indicates a forest, but also matter, substance, and is connected to the rabbinic *hiiuli,* prime matter.

Giacomo Leopardi talks about this in the *Zibaldone,* his collection of reflections on literature and philosophy. This original meaning of wood from the forest reminds us that in primitive societies wood was the construction material par excellence. After this, the Greek word shifted to imply any undifferentiated prime matter which, through the intervention of an organizing principle, gives rise to the multiplicity of the real world. In the word 'materia' there remains this trace of the female, of a passive, malleable element. In other Latin languages, like Spanish and Portuguese, wood is still named thus: *madera, madeira.*

If we relate this to the world of the peasant farmer, the *madre* (mother) becomes the stump of the plant, the innermost part of the tree from which new trunks, new shoots are generated. A matrix, a vegetal womb which produces the new wood, which is soft and workable. Matter as the source of the most docile and versatile of all materials, capable of adapting to any function.

This intimate link to generation is echoed in the myth of Hylas (resonating with the word ὕλη, *hyle*), the beautiful youth that Heracles fell head over heels in love with, making him his personal companion. Together they set sail with Jason and the Argonauts in search of the Golden Fleece. But during one break in the journey, Hylas, sent to draw water from the source of the river Pegae, encountered the nymph Dryope and her sisters, who in turn fell in love with the young man with the wonderful features. In order not to be separated from him, they led him into an underwater cave from which he never re-emerged.

The female divinities, the water and the submerged cave can only make us think of the creative principle. Of the body which in the darkness and humidity of its own belly generates, safeguards and nourishes life in embryo. That is how matter, a term paradoxically used to describe an inert, cold, inanimate component, acquires its *maternal* meaning from the first living matter with which we had a complex dialogue, for months: the female body which created us.

The rest of the story is simpler. Wood, the original primary material, lends its name to the more generic bodily substance which characterizes every distribution of mass in space. But its total, tangible concreteness, its body and substance, becomes the subject of speculation, a philosophical discipline which runs through the history of mankind, since even we humans, conscious beings, noble animate matter, as we define ourselves, are made of matter and, moreover, matter in its most fragile form.

In the specific case of us Sapiens, a specific species of anthropomorphic monkeys, things turn out to be even more complicated. Our being social animals is something more profound and integral than the simple fact that we live in organized groups. Our interaction with other members of the community, mediated through looks and language, bodily contact and exchange of food, acts of caring and emotional relationships, is a process fundamental to the growth of the individual. Effectively, we become truly human through the look and the exchange of emotions, interacting thus with other members of the social group.

The malleable and multiform brain of the new-born is formed in its relation to the world mediated through the adults who look after it, starting from its mother's look. The child, who looks into the eyes of the person feeding it, modifies its synapses on the basis of the reactions which are produced in the course of this relationship.

The drive to nourish and protect our little ones has a biological origin; it is a behaviour necessary to the reproduction of the species. We belong to the class of mammals, and this ingenious invention of evolution, whereby the females of our species are capable of nourishing for years little ones who would otherwise be incapable of surviving, has proved to be an enormous evolutionary advantage. This characteristic, which in its primordial forms developed around two hundred million years ago, is seen by some as the reason for the planetary success of mammals which in fact quickly occupied all the ecological niches left vacant by the disappearance of the great reptiles.

This also happened to the anthropomorphic apes from whom we have descended. That primigenial exchange of food between mother and child, that interaction of glances in a silent dialogue of protection and gratitude, is maybe at the root of every social connection and language that will develop over the millions of years to come. The astonishment at seeing nourishment for everyone – even for adults in the clan when the scarcity of food put at risk the survival of the group – gushing forth from the swollen breasts of mothers, can be found in the earliest artistic records of the Sapiens: dozens of prehistorical Venuses, all representatives of an archetype of abundance, goddess-mothers with swollen breasts and imposing buttocks.

But each person's material body, which plays such an important role in the creation of our first social relationships, the basis of our identity, is also an essential symbolic element at the other extremity of our existence, the moment of our death.

Mourning rituals and due respect for corpses

When an unexpected misfortune occurs, like the one that devastated my father's family, the whole of the little community who knew the victim relive the most ancient of traumas. A young man, a robust body full of life which, in an instant, slumps and becomes something inanimate.

The extreme precariousness of human existence already resonates in the words of Achilles, the greatest of the heroes of Ancient Greece who speaks of life thus: 'This thing which is so fragile and light; it lasts one instant and escapes so quickly through the mouth.' Deciding the fate of every mortal are the three Fates (Moirai), the daughters of Zeus and Ananke, the goddess of necessity, created out of the primordial Chaos together with Chronos, time, around whom she is wrapped like a serpent, to indicate an indissoluble bond.

When Atropos, the inflexible, severs the fine thread spun by Clotho and wound onto the spindle by Lachesis, there's no escape; even the strongest of the heroes collapses to the ground like a lifeless puppet, reduced to a heap of random limbs. In the terrible clashes which take place around the beautiful wall of Ilium, the robust bodies of the young heroes, who until a moment earlier seemed immortal, are transformed into shattered bones, disfigured faces, blood and guts; and with them the dreams, emotions and passions that drove them vanish.

Thousands of years later, despite the incredible progress that has led us to live more comfortable and much longer lives than those of our ancestors, an awareness of our intrinsic fragility is still with us. A fatal oversight, a tiny virus, a group of cells which goes mad during the process of infinite replications, a small blood vessel which suddenly gives way and even today we still have to come to terms with the trauma of someone's sudden death. Every day my father Giuliano's desperate cry: 'Babbo! Babbo! Breathe! Speak to me! Don't leave me!' as he held in his arms the tortured corpse of his own father, Guido the tailor, echoes around our roads or in the intensive care units of hospitals.

To relieve the trauma of loss, since the dawn of time, human communities have developed rituals for mourning and the burial of corpses. Paying respect to the poor tormented body, washing it and anointing it with perfumed essences, colouring it with ochre-red, enhancing with make-up features disfigured by death, adorning it with the most precious ornaments, cladding it in its favourite weapons, beautifying it with death masks made from the most precious metals, placing around it best-loved toys or the jewels that had enhanced its beauty, the insignia of authority or the humble tools of an artisan. And the monumental tombs, inscriptions, portraits, frescoed walls, hymns sung to accompany the deceased on their journey into the beyond.

At the centre of every burial ritual is the body of the deceased, which the most ancient of all tabus protects from the mangling which wild animals could inflict, were it to be abandoned.

Achilles, the strongest of the Achaeans, has just pierced Hector's throat with his lance. The Trojan hero just has time to say his final words while the most furious outburst of anger distorts the features of Peleus' son. Achilles pulls out the bloody lance from his enemy's throat and strips the body of the magnificent bronze armour which the Trojan had snatched from his friend Patroclus. He considers leaving the flesh of his dying enemy to be torn apart and devoured by dogs and birds, then, without hesitation, he pierces a large hole in Hector's feet to slip a rope through. He ties the young man's still warm corpse to his chariot and drives his horses at full gallop right up to the walls of the enemy city. Achilles butchers Hector's body in front of the Trojans who watch on in horror.

His enemy's body will lie, abandoned for days, near to Achilles' tent close by the huge, beached keels of the Achaean ships. Through divine intervention, no dog or bird will approach to relish the flesh; on the contrary, his wounds will heal, including those inflicted, out of contempt, by the Greeks on the defeated man's corpse. Not a single sign of putrefaction will profane the body of the hero who had died in battle.

The miracle will continue until, after twelve interminable days, Hector's elderly father Priam hastens in the middle of the night to his enemy's tent with a cart laden with riches to beg for the return of his son's mortal remains. The old man humbles himself in front of Achilles, clings to his knees, kisses his hands which are still red with his son's blood, just to have him back, even dead. He wants to pay due respect to the corpse, as is right and proper. And at this point, Achilles, the beast, capable of throttling someone without pity or burning twelve Trojan youths on Patroclus' pyre, yields to the old king's plea.

It is of such importance, to the ancient Greeks, to reaffirm the tabu of the inviolability of corpses, that this episode from the *Iliad* will become an ideal point of reference for all successive armistices. Even in the bloodiest of conflicts, there will be a moment when armies cease fighting to exchange the bodies of the fallen. It is impressive to note how this practice has survived up to our time; you simply have to look at the news coming from the Russia–Ukraine war. Even in the terrible conflicts of the twenty-first century, despite being fought with missile strikes and satellite technology, there comes a moment of pity, an ancient ritual in which weapons fall silent and soldiers load onto their shoulders the remains of their fallen comrades who had ended up in the hands of the enemy.

Still today the suffering felt from having to weep over a death without being able to honour the corpse is intolerable. This can be seen every time a flood or a tsunami makes it impossible to recover the bodies of some of the victims. It is an unimaginable suffering for relatives not to be able to place together in a coffin the poor remains of their loved ones destroyed by acid, as in some Mafia crimes, or thrown into the ocean from a military plane as happened to many of the Argentinian *desaparecidos*. When you want to inflict the most inhuman suffering, you commit murder and destroy the corpse or make it disappear forever, such that the relatives of the victims are denied even the consolation of crying and grieving while embracing what remains of their loved ones. Death transforms the bodies into a truly special material, impregnated with symbolic meanings, which we simply cannot do without.

The power of a body made inanimate can be even better understood by considering the mourning practices documented by ethnologists among groups of chimpanzees and bonobos. They just had to observe more closely what was happening in these small societies of anthropomorphic apes, who are very close to us genetically speaking. The death of a member of the troupe, particularly if they occupied a significant position in the social hierarchy, draws the whole clan into a whirlwind of extreme emotions in which laments and shouts, expressions of threat and submission alternate. The ritual can last for days on end and often leads to a refusal of food, a kind of collective fast, seasoned with forms of aggressive behaviour interspersed with demonstrations of comfort. In many cases the corpse is watched over, even at night, by individuals who touch it with light taps similar to caresses. The tiny corpses of little ones are cradled by their mothers, who hold them in their arms for days, delicately removing their fleas and keeping flies away from them.

The most intense reactions are triggered by sudden deaths, as if the group were incapable of absorbing the trauma of an unexpected loss. In this case too, relatives and close friends exhibit the most extreme behaviours. For chimpanzees and bonobos, just as for humans, the bodies of the deceased become powerful symbolic elements to reaffirm the unity of the group. Inanimate matter, which up until a moment before had belonged to a living body, acquires a fundamental importance even for species akin to our own.

This is what maybe gives rise to the most ancient burial rituals, already documented among the Neanderthal, the species which preceded the Sapiens by hundreds of thousands of years. The corpse of the deceased is at the centre of collective rites for processing grief, which sometimes even included forms of ritual cannibalism. The social function concludes with ceremonial burials in natural cavities or ones dug out purposely inside caves. The posture of the corpse, the ritual placing of stones or artifacts, the use of various kinds of pigments and decoration, are evidence that our most distant forebears carried out ceremonies which were destined to alleviate the trauma of loss.

Matter in the thoughts of the great thinkers

An awareness of the extreme fragility of all living forms emerges way back in the mists of time. And it's not surprising that that awareness gave rise to the earliest religious beliefs. In the Latin etymology of the word *religio*, many people find a binding value of sacred obligations and prohibitions; I prefer to see in it a narrative which bonds, holds together a community ripped apart by anxiety and suffering. Religion is born from an act of rebellion; the end of the material substance of which we are made quite simply cannot mean *our* end. Something of us must remain. In a material world dominated by the natural world, which follows temporal cycles that repeat indefinitely, the intolerable thought arises that nothing remains of the single individual.

The great material structures like rivers and mountains, the Earth and the Sun enjoy the absolute privilege of an eternal existence, as if they were made up of incorruptible matter. It's impossible that we, such special beings in so many other ways, are, by contrast, condemned to deterioration and death. Something of that substance which makes other things indifferent to corruption and to time must be hiding in our innermost structure too. It's inevitable that we should think that even humans are connected to the subtle fabric which lives for eternity. The consoling power of a belief that something of ourselves resists the destructive force of time, that we shall see our loved ones again and that their broken existence might be pieced back together again in its affections and its relationships, builds an invincible armour that increases human resilience in the face of the worst catastrophes.

This, in all probability, gives rise to the earliest beliefs in a beyond, to ceremonials preparing the deceased for life in the other world, to burials in the belly of mother-earth, in contact with the immortality of its most profound structure, capable of regenerating the individual through a new birth.

When you imagine a noble element, a fundamentally alien, irreducible substance, which animates the corruptible matter from which our bodies are made and which survives forever the destruction of death, then it's evident that the coordinates around which the bulk of philosophical reflection on matter will develop are already clearly defined. On one side, the idealization of this substance which will take the name of vital form or breath, of soul or spirit; on the other, the reduction of the body to dust, matter both quintessentially vile and malleable.

The most famous of stories is born from this: 'The Lord God formed man of the dust of the ground and breathed into his nostrils the breath of life and the man became a living soul.' The birth of Adam, the first man in the book of *Genesis*, makes this distinction explicit. Dust, the raw material which makes us similar to the animals, is responsible for the basest instincts which stir our existence: hunger, sex, violence. The malodorous dust out of which we are made gives rise to the drive to dominate and bully others and the blind necessity to reproduce ourselves as a species. The creative act of the divinity ennobles that substratum in which a tangle of bestial instincts continues to stir, sets its limits, and passes on the breath of eternity. The soul will be temporarily hosted in a body condemned to die and deteriorate. At the moment of its passing, it will split away to be united once again with the divinity which created it and enjoy being in its presence for all eternity.

I find it evocative that, to make the creation story incisive and unforgettable, the Bible uses the image of a God moulding humanity out of the dust of the earth. A divinity using the same technique developed by humans many thousands of years before the book of books was written.

The production of terracotta vases and utensils in fact marks a milestone in human history. People are creating, literally, something which didn't exist before, imagining its form and bringing it into being with the skilful use of the hands. I have often found myself enchanted when faced with the mastery with which the earliest terracotta artifacts were fashioned: receptacles for preserving water and food supplies. The extreme malleability of clay allows our distant ancestors to organize space differently, separating outside from inside, building a void, or the inside of the receptacle which will be filled with water or food reserves, and this can be transported, making hunting or gathering more efficient. In the biblical story the euphoria that accompanied the discovery of the enormous possibilities of this new technique seems to resonate still.

In *Genesis* we can already detect that same dominant note which runs through the thinking of the ancients: matter as a passive, chaotic, feminine element, in dialectic with the masculine element, the active principle, God or the Demiurge which gives it form and actually brings it to life.

The concept of matter as the womb generating everything, but also as an indistinct, shapeless, nebulous element is shared by the pre-Socratic thinkers who were investigating its composition and about whom we shall speak when we discuss the origins of atomism and of materialistic conceptions.

The great thinkers of antiquity are moving on the same wavelength. Plato considers matter to be a shady vessel for the forms which come from the divine world. It is the amorphous source of evil and corruption in the sensible world. In as much as it lacks form, it is unstable and thus incapable, in and of itself, of constructing a lasting and permanent equilibrium. In Aristotle, matter is corporeality, substance, the constituent element of things, which nonetheless remains indeterminate. We cannot know it fully because it is potential, still something amorphous, a passive component of the universe which exists in chaotic indeterminacy and can be transformed into anything.

As can be seen, the ancient thinkers remain prisoners of the old assumption that the noble, rational part of the human soul can have nothing to do with a birth from a woman's body, with matter which takes shape in the flesh and blood of another being, and a woman at that. It's no coincidence that Athena, the goddess of wisdom, is born fully armed from Zeus' head. Rationality, temperance which shuns blind ferocity, the goddess of balance and justice who finds every cruel act repugnant, can only be born from the mind of a man.

An assumption about the stability and persistence of the material universe

An awareness of the extreme transience of all biological forms, and of us human beings specifically, constitutes the starting point of the great religions and the first philosophical speculations. From the dawn of time, the illusion of imagining that something of us will survive has characterized human thought. But whereas here and there the doubt has surfaced as to whether it is really possible to escape the inexorable cycle of life and death, very few people will challenge the assumption that the material universe which surrounds us is enduring and stable.

A superior being which gives form to Chaos and transforms it into a Cosmos gives birth to a perfect universe, created in the image and likeness of the creator and, as such, destined to last forever. 'All things, among themselves, / possess an order; and this order is the form / that makes the universe like God,'* Beatrice explains to Dante in *The Divine Comedy*

* Quotation from Dante Alighieri, *The Divine Comedy*, translated by Allen Mandelbaum, Everyman, 1995.

(*Paradise*, canto 1, lines 103–105). It is irrelevant that in certain religions, like Christianity, and in certain schools of thought, like Stoicism, there is speculation about an end to the created world on the day of the Apocalypse or a succession of universes that are destroyed only to be generated anew. In its cycle of existence, whether longer or shorter, the material universe presents itself as a gigantic contraption, perfectly conceived and masterfully realized.

That is how one can extrapolate onto the whole universe an assumption which arises from our perspective. This, of course, now turns out to be an imaginary universe, since we extend to the whole, which is hugely vaster than the insignificant little corner that we know, the categories which are most familiar to us, ones which emerge from a point of view which is extremely limited in time and space.

Tiny and insignificant beings abrogate to themselves the right to construct a vision of everything. They do not know that they inhabit the lukewarm planet which occupies the third orbit around the Sun, an anonymous medium-sized star, more or less midway through its existence. They are unaware that it is only one of the roughly two hundred billion which inhabit our galaxy, which we call the Milky Way. Today, we are aware that that great barred spiral structure is certainly enormous, but very similar to an infinite number of others among the hundreds of billions which occupy the entire universe.

None of this could have been known by the great thinkers who more than two thousand years ago were reflecting on our condition. They could see wonderful celestial bodies rotating around them, whose motion was repeated, unchanged, since time immemorial. And that's how they conceived of a system of celestial spheres, perfectly organized, an ideal mechanism which controls an imperturbable equilibrium which has always existed and will always exist.

This assumption, concerning the eternity of the perfect matter of which celestial bodies are made by contrast with the transience of the base matter which makes up our bodies, will persist for centuries. When a new light appears in the skies, it will be abominable even to imagine that a star has died. Nobody of sound mind would even think that those perfect bodies too could be subject to the cycle of material existence. Thus, new stars will be the name given to those phenomena which today we know are nothing other than the extreme death rattle with which the existence of the most massive of the stars concludes.

This idea of a material universe in constant equilibrium will persist almost up until our time, until the early days of the twentieth century. Today, it seems astonishing when people observe that it was a view shared by every major scientist just a hundred or so years ago. Even the great changes in the vision of the world brought about by the scientific revolution of Copernicus, Kepler and Galileo were not sufficient to chip away at it.

For centuries, the Sun had been located at the centre of the system of planets and the Earth had irrevocably lost that central role which had deceived the ancient thinkers. Everything changed, but great scientists and ordinary citizens shared their assumption of a system in perfect equilibrium, both eternal and immutable. Astronomers had described and catalogued sunspots and had drawn the mountains and valleys that could be observed on the surface of the Moon in an abundance of detail, but nothing seemed to dent the idea of the special composition of celestial bodies. This idea of an ethereal material, which incorporates the possibility of moving indefinitely, even survived the discoveries of Isaac Newton and his proof that the motion of the planets derives from a force originating from the mass of the celestial bodies, with no need of special materials to produce gravitational pull. Rather, even the base *sub-lunar* substance which our Earth is made up of, manages to attract the apple which falls from the tree and keeps us firmly on the ground.

But the assumption whereby all of this is part of a material scenario which is eternal and immutable, part of a perfect equilibrium which has existed and will exist forever, will survive as far as Einstein, who will rebel until the last against the idea that everything had a birth, and will have to pencil in by hand, that is to say arbitrarily, a form of positive energy from the void, which pushes everything away from everything else, in order to avoid the material universe described by his equations of general relativity being unstable.

This book will help us to overcome many prejudices, but above all our shared discussion of the discoveries of contemporary physics regarding matter, its birth and its long evolution, will leave us gasping for breath. In the first place because we will come across wonderful concepts, discover ephemeral forms of matter which live apparently insignificant existences and, alongside those, in maybe the most unexpected corners, material forms which are so persistent as to be practically eternal. We shall be surprised to discover that we ourselves are in large measure made up of those. But we shall also touch with our own hands the intrinsic fragility of the most imposing material structures, a fragility which is generated by the most intimate, hidden mechanisms and which involve the behaviour of the elementary particles of which they are composed.

By analysing the great variety of surprising forms in which matter presents itself to the eyes of modern science, we shall have to abandon definitively the image of something corporeal, tangible, concrete. We shall discover forms of matter which are all around us, which play a decisive role in the macroscopic structure of our world, but which we cannot see or even touch, and about which, as of yet, we know nothing. We shall ask ourselves questions about the mechanisms which hold together the most compact of material bodies, capable of concentrating abnormal masses in incredibly small volumes, obeying unknown dynamics, which no law of physics has yet been able to explain.

We shall cross vast cosmic distances to investigate the most evanescent and delicate forms of matter, which scamper around the whole universe taking part in phenomena which are decisive to its evolution. We shall also go deep inside the most complex forms of matter, we shall seek to understand the mysteries hidden within biological matter, which we shall discover not to be base or amorphous at all. In it we shall see an astonishing organization at work, but one which is so connected to a series of immensely delicate equilibria as to be extremely fragile.

We are ready to explore the infinite facets of matter as narrated by contemporary science. We shall state that the assumption surrounding its stability and persistence, which we have believed for millennia, is a pure illusion. But at the same time, we shall realize that it's no easy thing defining what matter really is. Little by little as we approach its most intimate composition, we shall have to give up many of the certainties that we have inherited from the most ingenuous forms of materialism. And, as always, there will be plenty of surprises.

2

Atoms and the void

Let's begin with the most familiar form of matter, the one that we encounter in our daily lives, like the floor we're standing on or the computer keyboard that I'm using to write this book. In common parlance we consider an object that can be seen and touched, something we can interact with through our senses, to be material. This concept can be extended quite naturally to everything that we can see, even if it is too far away to be touched, such as the Moon, or that we can observe with the aid of instruments, such as telescopes which show us Saturn's rings.

Sight is a sense that we attribute great importance to, even if there are material bodies, like viruses, which are only visible if we have access to very powerful microscopes. On the other hand, there are things which we are in the habit of seeing even though they don't have their own material consistency, such as the shadow which falls on the sand when we're walking along the beach. In this area, we have to pay particular attention as our senses can deceive us. There are pathologies or altered states which cause hallucinations and there are artifacts which can be produced by the instruments that we use, not to mention well-known optical effects such as the Fata Morgana or mirages which make us see things when there is nothing material there. To be precise, there are very common material substances which we can neither see nor touch: for example, the perfume of a rose, which is undoubtedly something material, emanating from the beautiful petals of the flower and coming into contact with the olfactory sensors located in our nostrils.

In any case, whether it's a cloud or a mountain, the wings of a butterfly or a snowflake that we're talking about, we can describe the most intimate composition of these types of matter by means of a very limited number of elementary components, a handful of particles which, by binding together, are capable of constructing the various material structures that are most familiar to us.

The idea that it's possible to explain the organization of matter on the basis of a limited number of indivisible and indestructible particles dates from many years ago, emerging around the fifth century BCE in ancient Greece.

The birth of atomism

The modern town of Abdera is a small, anonymous community, situated in north-eastern Greece, on the fringes of the great flows of tourists which affect the country. Visiting it today, nobody could imagine that in the classical period it was an immensely flourishing city, rich in commerce and famous for its school of philosophy.

Archaeologists have rediscovered the remains of the ancient city on a promontory overlooking the sea, an enviable position, at the heart of a coastline rich in rivers and broad inlets. Tradition has it that the name derives from Abderus, a young hero who met a horrible end on those same beaches. A favourite of Hercules, the Demigod who was author of the twelve labours, Abderus was eaten alive by the man-eating mares of

Diomedes, the king of a barbaric population which lived in Thrace.

Ancient Abdera, founded by Greek settlers in the seventh century BCE, benefited from a double port, capable of offering safe anchorage, being well protected from violent winds. The city quickly became a required stopover for all the traffic crossing the Hellespont: hence its prosperity and the possibility for aristocratic families to endow it with imposing buildings for worship and public administration and to provide for the education of the young by attracting some of the greatest thinkers of the age.

Tradition points to Leucippus of Miletus as the founder of the city's important school of philosophy around the middle of the fifth century. But when speaking of the first great thinkers of the Greek world, it's always very difficult to distinguish clearly historical facts from traditions and legends. As often happens in these cases, sources disagree, so much so that even his city of origin is uncertain. Leucippus' thought has been reconstructed largely thanks to the writings of later thinkers, such as Aristotle who, around a century later, was challenging his theories. There are even doubts as to the historical figure himself; Epicurus, the great interpreter of the atomistic current of thought, maintained that Leucippus had never really existed. Nobody, however, doubts the existence of the character that tradition considers his favoured disciple: Democritus.

Scion of one of Abdera's wealthiest aristocratic families, and born there in 460 BCE, according to legend, he renounced the wealth and the prestigious roles which that wealth would have guaranteed him. Democritus is a prolific thinker, who ranges over all aspects of philosophy. It seems to be historically documented that, driven by his curiosity as to the customs and habits of different populations, he embarked on a journey to the East with a possible stay in Egypt. He does, however, spend most of his long life in Abdera. Tradition recounts that he will live more than a hundred years, a considerable age in absolute terms and distinctly extraordinary for the period. He has numerous disciples and writes works which cover the most varied areas of knowledge: from mathematics to ethics, from literature to music. To our great misfortune, the majority of his works have been lost. There remain around three hundred fragments which help us to reconstruct his thought, especially if integrated into the numerous quotations from Plato and Aristotle who contested his theses with great determination.

Democritus arrives at atomism from purely speculative bases, with a logical approach, typical of the philosophical arguments of the time. Like all Greek philosophers he abhors infinity, that non-number par excellence, the indefinite size which, as such, cannot be understood and thus contained; a process which goes on to infinity becomes a subversive entity, intolerable in as much as it conflicts irremediably with measure and rationality. And thus, the operation of subdividing any material component cannot continue indefinitely. Every material substance broken down and reduced to smaller and smaller fragments will have to be reduced to something that cannot be subdivided: an atom (from ἄτομος, *àtomos*, which means precisely that, indivisible). Matter cannot behave like Zeno's space, which can be broken down as much as you like. If things could be broken down to infinity, in the end there would be nothing left, and thus being would be made up of non-being. An absurd logic. And that's why we have to imagine something solid, impenetrable, uncreated and eternal, which limits the process.

Democritus' atom resembles, in many respects, Parmenides' 'Being'. The fundamental component of every material thing is a corpuscle, imperceptible, identical, perfect, lacking in any particular sensible qualities. Atoms are all made of the same substance, they only differ in terms of geometrical qualities, shape and size and are indestructible. Alongside atoms, Democritus theorizes another decisive component, the void, in which atoms can move, collide and band together in varying quantities, shapes and order, giving life to various elements. For the early atomists, therefore, reality is constructed by virtue of the combination of two ingredients: Being, fullness, that is atoms, and Non-being, the void.

Democritus confers on atoms the property of moving indefinitely and in every direction in the void, until they bond with other atoms to form material structures. Everything we see in the world occurs by virtue of this mechanism, which is both spontaneous and intrinsic to nature itself, requiring no external intervention. Even the soul is made up of atoms; they are spherical, thin, very light and capable of passing through any material body. The soul is mortal, because the atoms of which it is made up separate with the death of the body and the soul itself ends up dissolving.

The atomists' ideas spread rapidly like wildfire. In the following decades their theses were bitterly debated and refuted. Amongst their most authoritative critics, Plato and Aristotle are particularly notable.

Legend has it that Plato had personally ordered Democritus' books to be burned and had instructed the disciples of the Academy to avoid even speaking his name. It is a matter of fact that the philosopher of Abdera is never explicitly mentioned in Plato's writings, not even in those where he is openly contesting the atomists' arguments. The idea of a kind of matter which self-regulates and develops according to immanent laws was inconceivable in the Platonic system of thought. Not to mention the void and the thesis surrounding the mortality of the soul, subjects which were abominable to Plato.

No less radical was Aristotle's critique, which reveals a clear knowledge of the atomists' theses and challenges not only the concept of the void, but above all the results, in his word contradictory, of the attempts to explain transformations. The main target of Aristotle's critique is the theory of movement according to nature as a result of which atoms move perennially in the void, a theory which is irreconcilable both with the need for a Prime Mover (hypothesized by himself) and with his observations on the Earth's state of absolute rest. A second argument challenges the reduction in the difference between the elements, the composition of atoms, to geometrical qualities of the same linked to their form (spheres, octahedrons, dodecahedrons and so on) which would lead to logically inconsistent, contradictory results. But above all the description would not explain the mutation. The key argument, offered to demolish the atomists' theses, is generation. Without the presence of an incorporeal form, pre-existing in potential in the parents' seed, how could the reproduction of living beings occur?

The radical critiques of the atomists' arguments expounded by the two greatest exponents of Greek philosophical thinking deeply affected their destiny, but throughout the whole Hellenic period and, following that, during the Roman imperial age, they remained at the centre of the debate, thanks above all to the contributions of Epicurus and Lucretius.

Epicurus and Lucretius

The popularity of Epicurus in the modern world is the result of a misunderstanding. The term 'epicurean' has entered common usage as synonymous with a hedonistic attitude, which favours the enjoyment of material goods and considers the search for pleasure as the sole aim of existence. In fact, this was the argument of the Cyrenaic school from whom Epicurus distanced himself. He maintained that a wise man might attain pleasure only by freeing himself from the fear of death, and this can occur solely through a knowledge of the physical mechanisms which govern the world. The aim of philosophy is to allow human beings to attain spiritual peace, a perfect equilibrium, in which it becomes possible to do without anything superfluous. Epicurus invited his disciples to live in wisdom and moderation and to enjoy the pleasures of existence, keeping away from the excesses which would have enslaved them. But his detractors, to strengthen the argument, will end up taking his ethical positions to extremes and distorting them, which is the reason why the prejudice about the philosopher exalting earthly pleasures is still with us today.

Epicurus was born around 341 BCE on Samos, an Aegean island just off the coast of modern Turkey, and in 306 BCE transferred his school of philosophy to Athens. The choice of the ancient capital of culture testifies to his wish to strike at the very heart of Plato and Aristotle's philosophical constructs, towards which he immediately showed inflexible hostility.

His was the first of the great Hellenistic schools and stood out from tradition not least because it had its seat not in the squares or gymnasia where free men gathered, as with Socrates, Plato and Aristotle, but in his private house. It was a dwelling with fruit and vegetable plots and gardens, far from the city centre, in a part of town from which the surrounding countryside could be admired. But the most revolutionary feature of the 'garden school', as it came to be known, was that, for the first time, women were also admitted as disciples; in fact, a famous courtesan in search of redemption was welcomed, and even several slaves. Epicurus' school was one of the first institutions in which a form of democracy not unlike our contemporary one was practised.

Adopting and developing Democritus' atomism allows Epicurus to lay the foundations of a materialistic concept of the world, on which he will build a fully rounded philosophical approach. Reality is perfectly knowable to human intelligence. There is no need for any divine intervention to understand natural phenomena. Everything can be explained when the internal dynamics, which act within the material world, are identified.

The real is made up of bodies and of the void in which they move. Composite bodies are made up of atoms, are formed by bonding and are subject to corrosion and decay. This law also applies to the soul, which dies with the body. For that reason, there is no reason to fear death. Nothing will be felt ever again, since with the death of the body the soul too will dissolve.

Atoms are tiny, invisible and indivisible bodies. They live forever and constitute the essential ingredients needed for the construction of the material world. To refute Aristotle's argument, which challenged the complexity and the inconsistency of the mechanism through which the atomists explained the formation of the elements, Epicurus attributes to atoms differing forms, sizes and, above all, weights. In the void, atoms are in constant motion, they form a kind of continuous, gusting wind, hurtling downwards at enormous velocity.

To allow atoms to collide, to bond and to form composite bodies, Epicurus introduces declination, a natural, intrinsic tendency to sudden swerves, to casual deviations from the trajectory. This is his most important contribution to the atomistic theory, which will become highly popular when a fervent follower of his theories, in the Roman era, will put Epicureanism into verse, translating 'declination' with the Latin word *clinamen*.

Epicurus wrote around thirty books on widely differing subjects. Unfortunately, only extracts and fragments remain of nine of his works. But his theories have come down to us thanks to the success of Epicureanism in the Hellenistic age and to the works of his followers in the Roman era, and in particular Lucretius.

Titus Lucretius Caro, probably born in Pompeii around 95 BCE, left us the most complete compendium of Epicurus' thought in his verse work *De rerum natura* (*On the Nature of Things*). Information about Lucretius' life is scarce and fragmentary. It seems plausible that he came into contact with Epicureanism at the 'garden school' in Herculaneum, where one of the teachers was Philodemus of Gadara, a Greek philosopher who was guest of Caesar's father-in-law, the extremely wealthy Lucius Calpurnius Piso, owner of a magnificent villa on the coast.

In 79 BCE, when the eruption of Vesuvius destroyed Herculaneum, the Piso villa was buried under thirty metres of rubble. White-hot ash and lava invaded the house covering everything, including the dozens of statues of rare beauty which had made it a true wonder. The mantle of volcanic material also buried the library, turning the thousands of papyruses stored there into charred scrolls. Some 1,700 years after the eruption, when archaeologists started excavating, there emerged splendid, frescoed walls, great mosaics and numerous bronze statues, among the most beautiful ever created. In what had once been the library, they recovered hundreds of charred papyruses and their surprise was such that, for a long time, the villa became known as Villa dei Papiri. Nobody at the time – we are around the year 1750 – could have imagined that just a couple of hundred years later ingenious scientists would find a way of reading part of the content of the scrolls. Thanks to techniques of CAT scans, similar to those used in routine medical diagnosis, in fact, it was discovered that many of the scrolls were texts written by Philodemus of Gadara, the Epicurean philosopher who was protégé of the Piso family. And thus, a natural disaster, which might have wiped out that knowledge forever, in fact preserved it for millennia.

Lucretius' *On the Nature of Things* will suffer a similar fate to that of Philodemus' papyruses. The work in six books, which transcribes in magnificent verse a concept of the world which liberates human beings from the terror of death and of the Gods, is an immediate success. Cicero judges Lucretius to be a very great poet. Ovid considers him sublime; Tacitus notes the many who preferred him to Virgil.

But precisely because it spreads throughout the Roman upper classes, Epicureanism is destined to be heading for collision with Christianity, the new vision of the world which defines the final centuries of the Empire. In the second century AD,

Epicurus' ideas seem to have gained the upper hand; Emperor Hadrian grants the Epicureans considerable privileges, Marcus Aurelius establishes in Athens a chair of Epicureanism funded by the state. But this is also the period when the early Fathers of the Church begin their most combative refutations.

Clement of Alexandria, at the beginning of the third century AD, attacks Epicurus head on as blasphemous and a champion of atheism, in that claiming that men could achieve happiness on Earth by themselves meant, for a Christian, substituting oneself for God. Clement calls for war on all 'those who place atoms as a principle: poor men without faith, slaves to pleasure, who clad themselves in the name of philosophers'.

When Christianity becomes the official religion of the Empire, there will be little room left for Epicurus' arguments. A similar fate will befall *On the Nature of Things*. Lucretius and his most important work will end up buried under a mantle of hostility even thicker than the layers of volcanic ash which had buried the Pisoni villa. Slowly but inexorably, copies of this masterpiece will be destroyed and be lost or, quite simply, nobody will take the trouble to copy them out from the damaged originals. The *damnatio memoriae* (condemnation of memory) which will sweep over Lucretius and atomism might have wiped out their existence forever, had it not been for a fortuitous discovery made by a very lucky manuscript hunter.

An incredible discovery made by the secretary to an Antipope

In 1410, it seemed to Giovanni Francesco Poggio Bracciolini that he had heaven at his fingertips. At the age of thirty, he'd been chosen as the personal secretary to Baldassarre Cossa, recently elected Pope, who had taken the papal name John XXIII.

Born in Terranova, a small town near Arezzo, the son of a spice merchant who, among other things, had ended up heavily in debt, he had to abandon his ambition to study jurisprudence at Bologna University to become a notary. Deprived of the economic support of his family, he found employment in Florence as a copyist, a humble profession that he carried out, however, with such diligence as to rapidly merit the support of key intellectuals and politicians in the city. A voracious reader and student of all the works in Greek and Latin that he could lay his hands on, he attempted the great leap by transferring to Rome, supported by strong letters of recommendation. And there, things would have it that, suddenly, the wheel of fortune started to turn in his favour. The Pope's secretary was responsible for the pontiff's private correspondence and belonged to the tight circle of his most trusted collaborators.

For Poggio Bracciolini it was his life's dream come true; he, the son of a spice-seller, had been welcomed to Rome, into the papal residence, the heart of the institution around which all the political and economic power of the world revolved. But destiny had a terrible prank in store for him.

At the beginning of the fifteenth century, the papacy was living through one of its most bitter crises, which was destined to crush John XXIII. Baldassarre Cossa was not exactly what is usually considered to be a saint. Born on Procida, the small island on the Western edge of the Gulf of Naples, he was part of a wealthy but widely talked-about family; two of his brothers would be arrested and sentenced to death for acts of piracy and would manage to escape the noose only thanks to the intercession of their powerful brother Baldassarre, who was already chamberlain, or Minister of Finances, to Pope Boniface IX. As custodian of the Vatican treasure, he had set up a lucrative trade in indulgences; he would do deals with the wealthiest and most powerful families in Europe every time some particularly desirable ecclesiastical position had to be assigned. Rumours circulated that he had by no means been a stranger to the poisoning which had brought about the early demise of Alexander V. As for the vow of chastity which all ecclesiastics should have adhered to, gossips spoke of the uninterrupted coming and going in the papal apartments of widows, nuns and young wives. But he might have been forgiven for all this, had it not been for the hostility of the King of Naples.

That period was one of the most tormented in the history of the Church, which since 1378 had been torn apart by the Western Schism. For decades, the Christian community, profoundly divided, had elected two Popes, one in Rome and one in Avignon, each fiercely at war with the other. The era of the Popes and Antipopes reached its climax at the time of John XXIII when there were three contesting the title: as well as John, the Spaniard Pedro de Luna (Benedict XIII) and the Venetian Angelo Correr (Gregory XII). Ladislaus of Anjou, the King of Naples, intervened in the dispute, and, declaring John XXIII an Antipope, advanced on Rome with his army, forcing Cossa to flee and seek help from the Holy Roman Emperor, Sigismund of Luxembourg.

On the insistence of the Emperor, John XXIII summoned the Council of Constance, specifically to bring an end to the Western Schism. But, contrary to his expectations, Cossa himself was arrested, found guilty of simony, sodomy, incest, homicide and diverse other crimes, and deposed in 1415.

Dragged into the calamity of his protector's ill fortune, Poggio Bracciolini didn't lose heart, however, and decided to make the most of his stay in Germany to dust down his inexhaustible humanist passion. It was well known that in German monasteries dozens, maybe hundreds, of almost completely forgotten works of antiquity were being stored, because they weren't being systematically copied out by amanuenses. Bracciolini started visiting Swiss and German abbeys and monasteries in the area around Constance. In one of these, maybe at St Gallen, on the Swiss shore of Lake Constance, or at Fulda, an ancient abbey near Kassel, he made the discovery which would forever inscribe his name in history. He fell upon the manuscript of *On the Nature of Things*, a work whose title was certainly known to him, since it had been quoted by Ovid, Cicero and other greats of ancient literature. Bracciolini immediately had the manuscript copied by a scribe and sent the work to a friend in Florence to get other copies made. It was 1417 and, very soon after, Gutenberg's invention of the printing press with mobile letters would do the rest.

Very quickly Lucretius' philosophical poem, with its speculations about atoms and gods, about man's happiness and opposing the fear of death, would become a fundamental text which could not be ignored. Like all great books, the manuscript uncovered by Poggio Bracciolini would change the history of the world, influencing dozens of artists, scientists and philosophers.

Venus, Zephyrus and maggots in cheese

The beauty of Lucretius' poetry was a deciding factor in the dissemination of *On the Nature of Things* among intellectuals across the whole of Europe. And thus, along with the characteristic cadence of his splendid hexameters, the revolutionary ideas of Democritus and Epicurus spread everywhere.

Europe in the second half of the fifteenth century was traversed by strong contrasting views which would lead to Martin Luther's Protestant Reformation. But, despite the explicit arguments on the mortality of the soul and the perniciousness of religion, initially at least Lucretius' text did not attract the attention of the ecclesiastical authorities. Perhaps because the circulation of the text in Latin was limited, or maybe because the beauty of the poetry was such that many readers were fascinated by it, including not a few clerics, several high-ranking prelates and even a cardinal, Marcello Cervini, the future Pope Marcellus II. It is a fact that, when the Index of prohibited books was established in 1559, *On the Nature of Things* did not feature among the incriminated works. To get to the first banning of the book, we must wait until 1717, when it was translated into Italian and published in London.

This delay explains how the ideas of Democritus and Epicurus, so elegantly expounded in that poetry, will come to be known by hundreds of European intellectuals and will have a profound influence on philosophical debate and research activities of so many humanists and scientists.

Through a series of fortuitous circumstances, therefore, materialism is reborn and acquires new strength, thanks to an Antipope falling into disgrace. If John XXIII had not been arrested in Constance, Poggio Bracciolini, engaged as he was in his activities as papal secretary, would not have been able to dedicate so much time and energy to his passion for ancient manuscripts.

The good fortune of Lucretius' work was instant. At the end of the fifteenth century *On the Nature of Things* was inspiring not just poetry and literary texts, but also several of the masterpieces of our figurative art, such as Botticelli's *Primavera.*

The painting, completed around 1480, continues to attract the attention of visitors who flock to the Uffizi Gallery today. The canvas portrays nine figures from classical mythology surrounded by a grove of orange trees and laurel which opens onto a meadow of flowers. At the centre stands Venus, goddess of love and beauty, with Cupid, blindfolded, in the air above her, shooting his arrows. On the right, Zephyrus is embracing and possessing the nymph Chloris, who transforms into Flora the goddess of flowers. On the left, Mercury, slightly apart, almost absent from the scene, with the three Graces, dancing, lightly veiled, close by.

In this rebirth of nature, in the exultation of plants and flowers, the work celebrates Venus as mother-nature, the resplendent, benign goddess who populates the Earth with every living species, the source of prosperity and pleasure for humans and gods alike. It is the triumph of the goddess of beauty who, through the arts of pleasure, manages to placate Mars, the god of war. In the generative power of nature, which here takes centre stage, we find an echo of the hymn to the goddess of love with which Lucretius opens his poem:

> Lifegiving Venus [. . .] since through you the whole race of living creatures is conceived, born and gazes on the light of the sun [. . .] For as soon as the sight of a Spring day is revealed and the life-bringing breeze of the west wind is released and blows, the birds of the air are first to announce you and your arrival, o goddess, overpowered in the heart by your force. [. . .] since you alone guide the nature of things and without you nothing emerges into the sunlit shores of light.*

Hidden beneath the splendour of Botticelli's brushstrokes or in the lines of illustrious poets like Angelo Poliziano, images and ideas which faithfully reflect *On the Nature of Things* are appreciated by the intellectual elite of the period without coming up against obstacles of any kind on the part of the establishment. The discussion is wholly different when these turn into explicit philosophical theories and, above all, when they step beyond the boundary of scholarly debate, thus becoming the subject of public discussion.

This is what happens to Giordano Bruno, a Dominican friar, born in Nola close to Naples in 1548, who makes his theories known across Europe. He demonstrates his encyclopaedic knowledge in lectures and public debates and writes an impressive quantity of philosophical works, some in Italian, to spread his ideas among a much wider public.

Lucretius is a constant source and point of reference for Bruno's thought. *On the Nature of Things* is quoted

* Quotation from Lucretius, *On the Nature of Things*, translated by Walter Englert, Hackett Publishing, 2003.

explicitly both in his Italian dialogues and his Latin works. Bruno evokes Lucretius when he is considering the atom and the 'minimum' as original and constituent parts of matter and hypothesizes the spontaneous generation of all material species, including living ones, on the basis of the bonding of atoms.

Human beings are one life-form like many others, all of which originate from natural phenomena. There is no substantial distinction between material bodies which have a life of their own and those which are inert. They are both born from a limitless universe, full of innumerable worlds in which infinite possibilities can be realized. This vision of the world, totally opposed to the official geocentric one, very quickly attracted the attention of the Holy Inquisition, which in 1592 had Giordano Bruno imprisoned and subjected him to a lengthy trial.

It is interesting to note that, towards the end of the sixteenth century, alongside the educated materialism of intellectual circles, forms of religious materialism, albeit less sophisticated than the former, were also already circulating among a wider public, evidence of which comes from records of other trials by the Inquisition.

Domenico Scandella, known as Menocchio, was an obscure miller, married and the father of seven children, who lived in the village of Montereale, in Friuli. In 1584, when he was fifty, he was accused of heresy by the Holy Office. During his trial, the miller defended his ideas intelligently: 'When the body dies, the soul dies too, but the spirit remains and returns to God.' In his ingenuous statements we can see a re-emergence of the materialism of the pre-Socratics and Lucretius' theory of spontaneous generation. 'I said that . . . everything was chaos', the miller explains to the Inquisitor questioning him on the origins of the world, 'that is to say earth, air, water and fire all together; and that volume, going like this, formed a mass, just like cheese in milk, and in that mass maggots came about and they were the angels . . . and God was there as well, among that number of angels, also created from that mass at the same time.'

Menocchio's first trial led to a life sentence, but he was released a few years later. The Holy Office, however, continued to keep an eye on him, and in 1599 the Friulan miller was imprisoned once again, and this time sentenced to death. The sentence was carried out a short time before the same fate befell Giordano Bruno, burned alive in the Campo de' Fiori in Rome on 17 February 1600.

The parallel fates of the philosopher known throughout Europe and the anonymous Friulan miller allow us to understand just how deeply materialist theories had penetrated into sixteenth-century society.

The birth of modern science and atomism

The Lucretian concept according to which 'nature is free, no slave to masters proud; / That nature by herself all things performs / By her own will without the aid of gods'* found its most fertile terrain in the scientific world. In the early seventeenth century, science is fraught with the debate over the Copernican hypothesis which placed the Sun and not the Earth at the centre of the universe and finds in Galileo and Kepler the pillars upon which this new vision of the world will be built. With Galileo, the modern scientific method is born, implicitly embracing the Lucretian assumption that nature follows its own laws of transformation. At various points in his own work, the Pisan scientist seems to support the atomistic hypothesis and it's no coincidence that one of his most brilliant followers, Evangelista Torricelli, should dedicate much of his research to the production and study of a vacuum, an essential component in Lucretius' universe.

* Quotation from Lucretius, *On the Nature of Things*, translated by Walter Englert, Hackett Publishing, 2003.

Isaac Newton openly declared his admiration for Epicurus and embraced his atomistic theories which led him, among other things, to develop a corpuscular theory of light. The 'hard particles', as Newton defined the atoms which make up bodies, are imbued with forces by means of which they act on each other at a distance to produce varying chemical reactions. The form, durability and impenetrability of a material body depend on the characteristics already present in the particles which are its constituent parts. With Newton begins the long sequence of events which led scientists to develop modern atomism. Matter is made up of atoms and it is the forces acting between these particles which determine the nature of bodies.

In the second half of the seventeenth century, the experimental sciences – astronomy, physics, early modern chemistry, which for many years to come will be entangled with alchemy – become the principal instruments for investigating nature. To explain the results of their experiments, scientists are forced, gradually, to abandon every reference to Aristotelian principles. Subjected to chemical reactions, substances change their form, produce compounds, and radically modify their properties. Matter appears as much more variegated and complex, and, as a result of this, reference to the four classical elements – water, air, earth and fire – proves to be unacceptable. Everything, however, becomes much simpler if one accepts the atomistic hypothesis, and considers matter to be made up of primary particles imbued with specific qualities. Despite these proving to be invisible and imperceptible, and despite it being impossible to demonstrate their existence directly, the evidence gathered is impressive.

At the beginning of the nineteenth century, the English scientist John Dalton is responsible for the first modern version of atomic theory. A great scholar of the properties of gases, Dalton's interest in air and the dynamics of the terrestrial atmosphere led him to become a kind of precursor of modern meteorology. To explain the results and properties of certain well-known chemical reactions, he formulates his atomic theory and publishes his table of the atomic weights relative to six elements (hydrogen, oxygen, nitrogen, carbon, sulphur and phosphorous), all measured in relation to the weight of the hydrogen atom, taken as a point of reference and equal to one. Thanks to the atomic weights, chemical reactions become quantitatively predictable. This is the beginning of the great adventure which, in 1869, will lead to the periodic table of the elements, the work of the Russian scientist Dmitri Ivanovich Mendeleev.

If the various chemical elements are organized in increasing order of their atomic weight, we discover that some of their physical and chemical properties repeat at regular intervals and are, therefore, periodic. For example, the so-called alkaline metals – lithium, with an atomic weight of 7, sodium, 23, and potassium, 39 – have a shiny appearance, are soft and ductile and show very violent reactions with water. By classifying them on the basis of the periodicity of their atomic weight given that they are 16 units distant from each other, Mendeleev organizes them into a homogeneous group.

Following this approach, the Russian scientist constructs a table which contains all known elements. With regard to the places that he finds to be vacant in terms of the periodicity, he maintains that these are elements yet to be discovered and he can even predict their properties. Several years later, when gallium and germanium are discovered, and correspond fully to the identikit traced for them by Mendeleev, the success of the periodic table will be unanimous. Later it will become clear that, to avoid contradictions, the table of the elements will have to be organized in increasing order of the charge of the atomic nucleus, what today is called the atomic number.

In 1897, in an ironic twist of fate, at precisely the same time that Mendeleev's periodic table of the elements was experiencing its maximum success, Joseph Thomson, physicist and director of the Cavendish Laboratory in England, was discovering the electron.

The end of the nineteenth century thus marked the triumph of the atomistic model and, at the same time, the beginning of its crisis. It became immediately obvious, in fact, that electrons were to be found inside atoms; what for 2,500 years had been considered the indivisible element par excellence was, in fact, a compound state, had a structure and could be broken down into even more elementary components.

This marks the beginning of the great process which led twentieth-century physics to produce what today we call the Standard Model of particle physics.

3

They're just particles

Thomson's discovery of the electron marks the beginning of the experimental study of atomic structure. Scientists of the early twentieth century succeed in establishing techniques, which allow them to 'see the invisible'. Atoms remain bodies which are too small to be viewed directly, but experiments in physics are now able to subject various atomic models to scrutiny.

In order to understand what ordinary matter is made up of, the matter that forms a rock or a butterfly's wings, we need to get to grips with elementary particles. We can do this in stages, since the vast majority of the matter which surrounds us is made up of a very limited number of basic components.

Spelled out in the simplest manner, the issue can be formulated thus: matter is made up of particles which interact by exchanging other particles. That's it.

But how did it become possible to identify these particles? What does it mean that they interact by exchanging other particles? What techniques allow us to discover the most infinitesimal parts of matter?

The Dark Side of the Moon

To mix the materials she needed, his wife would use a metal container. The gadget could rotate thanks to a small electric motor capable of thoroughly blending the mixture which she would then transform into a new ceramic object. For days Roger had been looking for a solution to his problem and suddenly everything was clear. All he had to do was throw a handful of coins into the rotating mixer and record the sound that came out. By sampling it and mixing the whole thing with the sound of a cash register, he quickly obtained the 7/4 time-signature he wanted.

We have just witnessed the birth of *Money*, one of the most famous tracks in the history of rock. The piece, written by Roger Waters for Pink Floyd and released as a single in 1973, was a worldwide success, up there with *The Dark Side of the Moon*, which sold over fifty million copies.

In the early 1970s, an entire generation of young people were attempting, even in music, to track down that irreducible fault-line which would forever separate the present from the future, and which could already be detected in politics, economics, fashion and even interpersonal relationships.

With the release of *The Dark Side of the Moon*, rock would never be the same again. The fracture brought about by Pink Floyd was irreparable. Many avenues opened up, some completely unexplored, which would give rise to developments which up till then had seemed unthinkable.

We are talking about rock music, but the discussion could be opened out to include many other disciplines. There are moments when, suddenly, everything changes, when an equilibrium is broken which can never be restored, and one senses that unpredictable innovations will emerge from this moment of rupture.

I was recently struck by finding a Pink Floyd t-shirt among the favourites of my eldest grandchildren, Giuliano, eleven, and Elena, fourteen. I was even more astonished to realize that they were very familiar with the group that had marked my adolescence and that tracks written fifty years before, like *Money* or *Time*, were well known to them. Today's teenagers who listen to trap music and who look down on even the most famous rock singers, because they consider them to be too old, appreciate music produced even before their parents were born.

This happens because the break was so profound and so clean that, despite many decades having passed, a similarly significant leap hasn't been repeated. I smile when I think of how I would have reacted between 1965 and 1970, if an elderly gentleman had made me listen to the music that was in fashion when he was young, somewhere around the 1920s.

The new rules generated by that break must correspond to the tastes of the very young because they are still in use today. Something similar happened in physics in the early twentieth century. A brilliant experiment and a series of theories that were developed to understand the results, revolutionized the way of looking at matter and at the world. But still today, more than a century later, we are following the same path, using those same rules. For now, we haven't succeeded in finding anything more effective or productive in terms of our knowledge of the most intimate structure of matter.

Particle hunters

In 1908, Lord Ernest Rutherford developed the experiment that would open up the path to modern particle physics. The atoms of an ultra-thin gold foil were exposed to a flow of other particles, with a positive charge and high energy. Passing through the material, a fraction of those particles changed direction and the apparatus allowed him to measure how far they had gone off course.

To his great surprise, Rutherford noted that they often underwent significant deviations to the extent that, in several rare cases, they even 'bounced' back. The only possible explanation for this strange behaviour was that the atoms were composed of a tiny central nucleus, in which all the positive electrical charge and all the mass were concentrated, surrounded by an evanescent cloud of electrons, with a negative charge.

The use of particles to probe the most intimate structure of matter is the fundamental technique that we still use today. When, in 1912, scientists discovered that we are continuously traversed by a flow of penetrating particles coming from deep space, the radioactive sources used by Lord Rutherford were soon replaced by cosmic rays. Then, from the 1930s onwards, accelerators came into play and these machines would dominate the scene for the whole second half of the twentieth century. The result of this work, which lasted more than a century, is the Standard Model of fundamental interactions, which is the most advanced stage reached today in this long search for elementary constituents.

To understand how our work as particle hunters functions, we need to start with some considerations about the dimensions of the tiny objects that we wish to study.

A fine grain of sand is one of the smallest things that we can clearly see. Its dimensions are roughly one tenth of a millimetre (using the notation to powers of ten, which will prove to be particularly convenient, 10^{-1} mm, which means one millimetre divided by ten, or 10^{-4} m, given that 1 mm is one thousandth of a metre, that is 10^{-3} m).

Anything smaller than this does not allow our eyes to distinguish any details. For example, if we prick our finger and a tiny drop of blood comes out, however hard we try, we won't be able to see the red corpuscles, the cells which give our blood its characteristic colour. But if we use a good microscope, it will be possible to see the erythrocytes, as they are known in scientific parlance, very clearly. By magnifying the sample hundreds of times, they will appear to us as little biconcave discs, and we are able to measure their diameter which will be around 7 thousandths of a millimetre or millionths of a metre, or as we say, micrometres, that is 7×10^{-6} m.

The sadly only too familiar Sars-Cov-2 virus is much smaller than a red corpuscle. Its dimensions are somewhere between 60 and 140 nanometres, i.e. billionths of a metre, or thousandths of micrometres, $(60–140) \times 10^{-9}$ m, using once again the powers of ten. Even the most powerful optical microscopes would not be capable of identifying one of these viruses. They are so small that light comes up against its intrinsic limits. And it's not a question of having a more powerful microscope, capable of providing greater magnification, because here the wavelength comes into play, that property of light and of all forms of radiation which can be described as a wave. The minimum wavelength of visible light is around 400 nanometres; this means that with optical microscopes, short of having recourse to extremely sophisticated techniques or ones which use quantum phenomena, we will not be able to resolve smaller structures. The light will simply confuse them, it won't distinguish details, and will show us an unfocused, rough image which will prevent us from understanding what is happening beneath that scale.

To overcome this barrier, we shall need to have recourse to forms of radiation which are much more penetrative than visible light. The higher the energy of radiation, the shorter its wavelength turns out to be, and this translates into an ability to resolve details of smaller and smaller dimensions. This is the principle exploited by electron microscopes, instruments which illuminate the objects being examined with beams of accelerated electrons. The wavelength of this 'light' created by electrons can easily reach one nanometre, 10^{-9} m, as a result of which a rapid reconstruction of the images of the virus responsible for Covid was child's play.

When we go below the scale of nanometres, we enter the world of molecules and atoms. Today we are capable of viewing them by taking advantage of a range of techniques: advanced electron microscopes, scanning tunnelling or atomic force microscopes. The latter are special devices which use a very sharp probe to scan the ultra-polished surface of the specimen under examination; the apparatus makes it possible to record the infinitesimal deflections in the probe when it approaches an atom and thus to reconstruct an image of the atomic structure of the sample.

With appropriate adjustments, it is possible to view the smallest of the atoms, hydrogen, which has a dimension of one tenth of a nanometre, that is 10^{-10} m.

Maybe now we can better understand the difficulties faced by physicists at the beginning of the twentieth century and Rutherford's genius in setting up his experiment, given that the atom is a damnably small object which is intrinsically invisible if you try to view it with ordinary light.

Rutherford's breakthrough idea was to use alpha rays as 'light'; alpha rays are a form of high energy radiation produced by the natural radioactivity of several unstable elements, which he himself had discovered several years before. Today we know that alpha rays are ionized helium nuclei, that is helium atoms with their electrons removed: thus, very heavy charged particles which interact violently with matter. To avoid them being completely absorbed, Rutherford chose to 'illuminate' a slender gold foil with alpha rays. The precious metal is one of the most ductile substances known to man and leaves of infinitesimal thickness can be extracted from it.

The result left no room for doubt. The vast majority of alpha particles passed through the gold foil unscathed; only a tiny fraction were deflected at wide angles and a few actually bounced back. The only plausible explanation was that the positive charge of the atoms was not distributed evenly inside the volume but was concentrated in a tiny region at the centre of the atom, its nucleus.

Rutherford's experiment consigned Thomson's proposal, the so-called 'plum pudding atomic model', with the positive charge distributed throughout the inside of the little sphere, just like the dried fruit in the traditional Christmas dessert, to the archives. Rutherford demonstrated that the whole mass and the positive charge of atoms was concentrated in the nucleus, whose dimensions were therefore ten thousand times smaller than those of the atom. The electrons, which weigh two thousand times less than a proton, only marginally contribute to the mass of the atom.

In the early years of the twentieth century, it was therefore realized that the atom – that 'indivisible' object par excellence – has, in fact, an internal structure. If it is imagined as a small sphere 10^{-10} m in diameter, its central nucleus turns out to be concentrated in a tiny region of 10^{-14} m. If we consider the simplest atom, hydrogen, formed by a single proton around which a single electron is orbiting, its nucleus is therefore 1 femtometre, that is 10^{-15} m. As a result, to honour the memory of the great scientist Ernesto Fermi, this fundamental unit of measurement, which defines the dimensions of a proton, will be known as 1 fermi, abbreviated to fm, the same as femtometres.

And so, it turns out that the atom is largely made up of void. If we expanded a hydrogen atom to the dimensions of a football stadium, the proton which forms its central nucleus would be as large as an ant placed at the centre of the pitch, whilst the electron would be rotating around it on the highest of the terraces.

With the discovery of the internal structure of atoms, it was then only natural for scientists in the first half of the twentieth century to ask themselves what atomic nuclei were made of and above all what held them together.

Particles which bond with other particles

It was Rutherford himself who coined the term 'proton', borrowing it from the Greek superlative πρῶτον (*proton*), which signified the first, that which preceded everything else. It was a fortunate intuition which would later be confirmed by contemporary science, even if at the time it was thought that the fundamental constituent, the true 'atom' of matter had been found.

It was in 1919 that Rutherford succeeded, quite by chance, in producing the first nuclear reaction. By bombarding nitrogen nuclei with the same alpha rays he had used to attempt to understand the structure of the atom, he discovered that ionized oxygen and hydrogen nuclei, that is to say nuclei without electrons, just protons, were being emitted. It was therefore a natural step to imagine atomic nuclei made up of positively charged protons, with a cloud of negatively charged electrons orbiting around them, making everything electrically neutral.

While progress was being made in the understanding of the most intimate structure of matter, new problems arose, however, and questions emerged to which it proved impossible to find answers; how do protons manage to stay together in the nucleus? The laws of electromagnetism were well known: protons, endowed with the same positive charge, repel each other with a terrifying force if we try to confine them in those very reduced spaces. The *vexata quaestio*, which had tormented the first Greek thinkers, advocates of atomism, re-emerged in new forms: if protons are atoms, how do they bond with each other? What binds them together?

And then there was an underlying problem which concerned electrons. In their orbit around the nucleus, they move with accelerated motion and the laws of electromagnetism do not admit exceptions; when a charge moves in a circular motion it must necessarily emit photons and thus lose energy by irradiation. But if electrons lost energy, the dimensions of their orbits would have to reduce progressively to the point of making them 'fall' into the nucleus. At this point they would join with the protons, neutralizing their charge and the atom would collapse. So, with that atomic model, matter – all matter made up of atoms – would disappear immediately right from the earliest moments. Obviously, there was something massive behind all of this which the scientists of the period were missing.

The situation became even more complicated in 1932, when the English physicist James Chadwick discovered the neutron. Once again this had Rutherford's fingerprints all over it. It had been he who, in 1920, had hypothesized that in the atomic nucleus, as well as protons, there must also have been other neutral particles, neutrons, similar in every respect to protons except for the fact that they had no electrical charge, and that they must have been slightly heavier. In this way, it was possible to explain the differences in atomic weight between the various elements with the same atomic number, that is to say the same electrical charge in the nucleus.

Chadwick set to work and was successful. When he decided to bombard a disc of beryllium with alpha particles, he proved that the particles emitted had all the properties of a neutron. At this point, using protons, neutrons and electrons as his essential ingredients, it was possible to offer a convincing explanation of Mendeleev's periodic table of the elements. The simplest element, with atomic number 1, was hydrogen, formed of one proton and one electron. With atomic number 2, you got helium whose properties could be explained by assuming two electrons in orbit around a pair of protons connected to two neutrons, inside the nucleus. Then you moved on to lithium, atomic number 3, three protons and four neutrons in the nucleus and three electrons orbiting around and so on.

With the discovery of the neutron, however, understanding what bound atomic nuclei together became even more complicated. It now became necessary to explain how protons and neutrons managed to stay together in such a tiny portion of space. A new force had to be found, one that was much more powerful than electromagnetic repulsion which kept even neutrons, that is particles with no charge, squashed inside the nucleus. And it had to be a force which died immediately, as soon as it moved away from the nucleus itself. The evidence that effectively the most powerful of all forces acted on the nuclear scale would be gathered very soon, but we would have to wait decades to fully understand what was going on inside atomic nuclei.

Making the question even more complex was the discovery of a wholly new force, which acted only in the sub-atomic world: *weak nuclear force.* For many years, nobody had found an explanation for the radioactive decay of several unstable elements, which turned into other elements producing beta rays, and thus emitting electrons. Until, that is, Enrico Fermi, then a young physicist, put forward his theory: behind the beta decay lurks a new fundamental interaction. It had eluded us thus far simply because it is extremely weak, a hundred thousand times weaker than the electromagnetic force, and its kingdom is tiny, because it is confined within the infinitesimal world of nuclear distances. It is too weak to manage to hold matter together, but it plays an important role in transforming it, in making it decay.

In proposing his revolutionary theory, Fermi glimpsed a possible analogy between weak force and electromagnetic force, thereby opening the way to the third great unification of the forces, between electromagnetic and weak forces which will be proved in the second half of the twentieth century. When, after some initial scepticism, the entire scientific community took the so-called Fermi interaction very seriously, physicists found themselves facing a double challenge: trying, on the one hand, to understand the monstrous force which held the nucleus together, and, on the other, to discover what gave rise to this second force which caused several components to decay.

It will take almost fifty years for the particles responsible for these new interactions to be identified. To achieve this result, physicists will first use cosmic rays and then develop the powerful particle accelerators which we still use today.

Only in the second half of the twentieth century will it be possible to understand fully what atoms are made up of, and the dynamics of the new force which manages to hold nuclei together, but in the meantime, there will be surprises a plenty. Because it will be discovered that protons and neutrons are not elementary particles either, but in their turn are made up of other ingredients, which are even smaller and very bizarre.

The irresistible force of Eros

> Verily at the first Chaos came to be, / but next wide-bosomed Earth, the ever-sure foundations / of all the deathless ones who hold the peaks of snowy Olympus, / and dim Tartarus in the depth of the wide-pathed Earth, / and Eros (Love), fairest among the deathless gods, / who unnerves the limbs and overcomes the mind / and wise counsels of all gods and men within them. (Hesiod, *Theogony*, ll. 113–119)*

It is fascinating that in one of the first great poetic works of the Greek world, Hesiod's *Theogony*, written around the eighth century BCE, Eros appears as a primordial divinity. The god of love precedes not only Zeus and the best-known gods of Olympus, but even black Night, Uranus the Heavens or the immense Ocean. With the power of Eros, Hesiod explains the attraction between the various divinities, the power of love which produces fusion and mingling and creates new entities. Eros represents something irresistible, which acts

* Quotation from Hesiod, *Theogony*, translated by Glenn W. Most, Harvard University Press, 2018.

remotely and turns the destinies of men, and even gods, upside down.

Making a leap of over two thousand years, we can see a connection between these lines and the famous hendecasyllable which concludes the final canto of Dante's *Paradiso* and with that the whole *Divine Comedy*: 'The Love that moves the sun and the other stars.'*

Here the majesty of the poetry is coupled with philosophical and theological precision. To describe the perfect, eternal and immutable material mechanism which animates the entire universe, that wonderful system of concentric spheres that Dante has barely finished admiring, and managing even to contemplate God, the Supreme Poet uses the word love, something radiant which emanates from the immobile prime mover.

No theory developed by modern science can compete in terms of beauty of form with the rigour and splendour of the poetry, but it is intriguing to note the extent to which this idea of remote action has had a disruptive impact on the history of physics, since people first started to try to explain the interactions between material bodies.

Isaac Newton was one of the first modern scientists to formulate a theory of forces which implied remote action. Until 1687, the year in which the great English scholar published his most famous book, *Principia*, scientists' thinking was dominated by the assumption that for a force to be exerted there had to be contact between two bodies: the hand which throws the stone, the horse which pulls the carriage, the weight which lengthens the spring attached to the ceiling.

By analysing the motion of the planets and seeking to explain the laws which govern their orbits, Newton reached the conclusion that between the celestial bodies a force of attraction is

---

* Quotation from Dante Alighieri, *The Divine Comedy*, translated by Allen Mandelbaum, Everyman, 1995.

at work which is developed remotely. It is the famous universal law of gravitation: a force of attraction between two bodies which is proportional to the product of their masses and inversely proportional to the square of the distance between their centres. For the first time in physics a concept comes into play which had previously seemed to be confined to the superstitious world of astrology: that heavenly bodies exercise an influence over phenomena at a great distance. Some maintain that, in fact, it was precisely because he was a staunch advocate of astrology that Newton was able to formulate his most famous law. Paradoxically, he would succeed in discovering a fundamental law of physics by starting from a totally irrational and erroneous assumption.

Newton's discovery is still important because it deals with the first unification of two natural forces, which until then had been considered different and separate. The apocryphal anecdote of the apple falling on his head conceals an extraordinary scientific result: the force which detaches the apple from the branch and makes it fall to the ground has the same origin as the force which keeps the Moon in its orbit around the Earth. Terrestrial gravity and celestial gravitation are two different ways of seeing the same force. Moreover, with Newton, a new chapter begins, heralding later developments. The source of this force is mass; the matter which a body is made up of has its own property, it is charged with gravitational attraction, it is capable of attracting any other body in the universe, it interacts with any other mass.

This modern approach will return two hundred years later, towards the end of the nineteenth century, when scientists attempt to understand electrical and magnetic phenomena. The laws of electromagnetism, which will be developed by Coulomb, Faraday, Maxwell and others, will lead to a second great unification of forces. In this case, it will be understood that electrical and magnetic phenomena, which had been thought to be due to independent, separate natural forces, are in fact two different manifestations of the same interaction: electromagnetism. 'Electrical charge' will be the name given to the source of these new forces which act remotely, and it will be discovered that in nature it appears in two distinct species, conventionally known as 'positive' and 'negative'. Charges of the same sign will repel each other, whilst opposite charges will attract.

To better explain the global effects produced by static or moving charges, the concept of force fields will be developed, associating these new properties with space as a whole. Finally, it will be discovered that electromagnetic fields can propagate oscillations. It will prove relatively simple to produce waves and transmit them remotely in order to communicate, and also to discover that light and optical phenomena are nothing other than manifestations of this new interaction.

The formulation of the laws of electromagnetism will lead us directly to modern physics because it will be precisely by asking ourselves questions about some of these paradoxes that special relativity and quantum mechanics will be born. The world of infinitesimal atomic distances, through which the electrons which orbit around the nucleus move, is the undisputed world of this new way of looking at matter.

For an electron, moving at velocities close to the speed of light is child's play. Since it has an electrical charge, it takes very little to accelerate it; it just needs to be kept in a vacuum and subjected to a strong electrical field and it will suddenly shoot off at formidable speeds. The laws of physics that govern the infinitely small generate in such tiny, light objects behaviours that are so different from those we are used to as to appear at the very least bizarre. The state of a system, space and time, mass and energy, everything becomes crazy in the world of atomic or sub-atomic particles.

The principal idea of quantum mechanics is that elementary particles cannot exchange energy in the form of a continuous flow, but only in packets, small discrete quantities which we call *quanta*, a term which comes from Latin, and which means quantity. For this reason, electrons can quietly rotate around the nuclei, following their circular orbit or thereabouts. They cannot emit energy by radiation, since it would be below the allowed minimum threshold, and thus they remain effortlessly in orbit and the matter doesn't decompose. It seems a negligible detail, but in fact this is precisely what explains the stability of atoms and, in good measure, the stability of the matter which the universe is composed of.

In Greek myth, it's the arrows shot remotely by Eros which produce an irresistible attraction between humans as between gods. In the world of contemporary physics, it is quanta flying from one corner of the universe to another which bond, indissolubly, all kinds of particles to each other.

The realm of the shyest and most self-effacing of the particles

Every force has an associated particle which transports it, the *quantum* of interaction. In the case of electromagnetism, this is the *photon*, the light particle. The action of a force is explained in a wholly new way. For example, the elastic repulsion between two electrons, which repel each other because they both have a negative charge, is described thus: one of the two emits a photon and is sent off in one direction, the other absorbs the photon and is sent off in the opposite direction. As a result of their interaction, the two electrons move away from each other.

Everything adds up, even if, obviously, we have violated one of the fundamental principles of physics: the conservation of energy. The photon that has been emitted transports a certain quantity of energy, and as it travels towards the electron that will absorb it, the system will contain, for a certain period of time, more energy than the initial amount. In classic mechanics that would not be possible, but quantum mechanics follows different laws and considers this violation, as long as it happens only for a very limited amount of time, governed by the uncertainty principle.

The iron rule which governs the world of quanta was expounded by the German physicist Werner Heisenberg in a 1927 article and since then nobody has managed to find a process which violates it. It has become a kind of axiom of quantum mechanics, intimately connected to the continuous fluctuation of the quantum states which characterize the world of small dimensions. In this specific case, the uncertainty principle tells us that the greater the energy transported, and thus the more serious the violation of the conservation of energy, the shorter the period of time in which this happens must be. From this we derive a close relationship between the radius of action of the force and the mass of the quantum which transports it.

Since no quantum of force can carry less energy than that which corresponds to its mass, that is how, by means of the photon, which has zero mass, electromagnetic interaction can expand into infinity. Any charged particle interacts with all the other charged particles of the whole universe, wherever they are distributed.

Using similar arguments for the weak interaction, it transpires, on the other hand, that the quantum of this interaction, whose effects are confined to sub-nuclear distances, must have a very large mass. In the mid-1960s, when a coherent theory of the weak interaction was being elaborated, it turned out that the carriers of this interaction, called W and Z, must have had excessive masses, respectively equal to 80 and 90 times those of a hydrogen atom.

For this reason, the weak force dies well before reaching the confines of the atomic nucleus. Given that it is confined within such tiny dimensions, it is not surprising that it took humanity thousands of years to become aware of its existence.

Thanks to the development of large particle accelerators, it was possible to reconstruct what was happening inside the nuclei. The idea that there had to be a 'strong' force, at least one hundred times more violent than the electromagnetic one, had been established for some time. Among the first to propose this we find Heisenberg again and Ettore Majorana, Enrico Fermi's outstandingly brilliant young collaborator, but the real response could only come from experiments. To probe sub-nuclear distances and manage to really 'see' what was happening inside the nuclei, particle accelerators were needed.

Electrons are so light that accelerating them to velocities indistinguishable from the speed of light turns out to be fairly straightforward. The trick is to use positive electrical fields and to make them run inside an ultra-high vacuum to avoid any kind of collision, which would make them lose their energy. The most effective solution is to use doughnut-shaped machines inside which electrons move in circular orbits. In this way, they can pass through the same region of acceleration many times and gain energy each time they pass. Appropriately placed magnetic fields are used to bend their path and force them to maintain circular orbits, so as finally to be able to extract them and lead them into collision with the target. All the way round, sensors are installed, capable of revealing the kind of particles which emerge from the collision and their properties.

The secret is to exploit the relativistic increase of the mass; the closer things get to the speed of light, the more the acceleration undergone by the electron makes its velocity increase marginally and on the other hand increases its mass. This is another of the effects of special relativity. The energy ceded to the electron by the electromagnetic field is no longer capable of increasing the velocity, since c remains a limit value and thus goes to increase the mass of the object.

Highly energetic electrons, which travel at velocities close to the speed of light, are capable of penetrating the nucleus with ease. This is what scientists managed to achieve from the 1960s onwards and since then a detailed understanding of their structure has become possible. As had happened with Rutherford, by studying the angular distribution of electrons which emerged from the collision, it was discovered that the mass of protons and neutrons was not evenly distributed throughout the volume but was concentrated in a few points. Basically, protons and neutrons too were formed from other elementary particles, which behaved in such a bizarre way as to merit the equally bizarre name *quark*.

Thanks to research carried out in the second half of the twentieth century, the properties of quarks were identified, and people were able to understand every detail of the behaviour of the strong force and the weak force with which they interacted among themselves.

Quarks have a fractional electric charge, 1/3 or 2/3, and are also endowed with a *strong charge* and a *weak charge*. They are the only particles capable of interacting among themselves by means of all the forces of nature. The strong force is carried by *gluons*, from the English *glue*, the literal translation of which could be 'stickies', to indicate particles capable of binding quarks together. By exchanging gluons, quarks undergo a violent attraction capable of overcoming any electrostatic repulsion.

Inside the proton, the quarks jump around here and there like very light wrens, flying in an instant from one point in the small volume to another. They are as agile as crazy wild cats, despite being limed in the super-strong glue with which they play hide and seek.

The strong force has very particular properties. The gluons have zero mass and are electrically neutral, but they too are endowed with a strong charge and can thus also interact with other gluons, even with themselves, making everything somewhat complicated. Having zero mass, like photons, their action radius should in theory be infinite, but different from the quanta of electromagnetism, gluons emitted by a quark bond strongly with everything that surrounds them, as long as it has a strong charge. From this a pattern of behaviour emerges which is quite different from the usual one. With electromagnetism, the distance between two charges reduces the force with which they attract or repel. With the strong force the opposite happens; as the distance between two quarks increases, the strong force increases, as would happen if an extremely robust spring were operating between them.

For this reason, quarks cannot live alone, but must always be bound to other quarks. When we try to separate them, the strong force field becomes concentrated into a kind of tube which lengthens as it tries to maintain the link at all costs. The energy of the field gradually increases as the distance grows and, when the situation becomes untenable and the link breaks, the energy which is released transforms into other quarks, which bind onto the ones that were attempting to be liberated. In conclusion, the action radius of the strong force cannot exceed 10^{-15} m, the dimensions of a proton.

Quarks, the elementary components of matter, behave as if they were very self-effacing, extremely shy particles, who in no circumstances agree to be seen *naked*. When you manage to shatter a proton, in fact, free quarks are never seen emerging from the collision; barely have they broken their bonds with the other quarks which made up the proton, than they *get dressed again*. The strong field, disturbed by the collision, has generated new quarks which have bonded to the previous ones and reconstituted other particles.

The eventual possibility that quarks and gluons might in their turn have an internal structure is a question which is still open. However, studies carried out up until now would seem to indicate that we are dealing with point particles. Their dimensions appear to be smaller than 10^{-19} m.

Five small phenomena

The most intimate structure of matter can be described fairly simply thanks to a subdivision of roles between the fundamental forces. Gravity, or the force which dominates matter on a vast scale, the force which holds stars, planets and great galaxies together, is so weak that, when its value is calculated in the case of microscopic objects, it turns out to be totally irrelevant. One can describe what happens inside atoms or their nuclei ignoring gravitational attraction completely.

Atoms and molecules are held together and bind to each other above all thanks to electromagnetic interaction. The radius of action of nuclear force is too small to play a role on objects of these dimensions. Nuclei and their components, protons and neutrons, are rather held together by the strong force emanating from the quarks which make them up.

This hierarchical organization of matter allows us to obtain important information about the bodies which surround us, even if we overlook many details of their internal structure. For example, in the same way that quarks are irrelevant in understanding the three-dimensional structure of a virus, so atoms are a pointless distraction if we are having to calculate at what angle and at what speed a basketball must be thrown to score a basket.

Ordinary matter, on the microscopic level, is made up of very few elementary components. Three particles of matter – the electron and the two lightest quarks, *up* and *down* – and two particles which carry the forces: the photon, which transports the electromagnetic force and acts on all three of the particles of matter, and the gluon, which transmits the strong force and interacts with the quarks, endowed with a strong charge, whilst ignoring the electrons which, on the contrary, have none.

As can be seen, with very few ingredients we are already in a position to explain a vast quantity of material structures. By combining two *up quarks*, each with a charge of +2/3, and one *down quark*, which has a charge of −1/3, we obtain a proton, with a charge of +1. If, on the other hand, two down quarks and one up quark are combined, then we get a neutron, which has a zero charge. The quarks that compose neutrons and protons are continually exchanging gluons, and the attraction that derives from this exceeds by quite some measure the electrostatic repulsion between quarks of the same sign.

The dimensions of protons and neutrons, 10^{-15} m, are enormous by comparison with those of quarks, which are at least ten thousand times smaller. And so, history seems to be repeating itself; even the components of atomic nuclei, so compact in appearance, turn out to be made up mostly of void, but it's a void dense with strong force, by far and away the most potent of all the glues. The three minuscule quarks move around continuously in a stormy sea of gluons.

The strong force between quarks and gluons is so intense that, when protons or neutrons are very close to each other, all it takes is a small gust from the storm roaring inside them to keep them tightly glued to each other. What keeps nuclei together is a marginal residue of the strong force which keeps quarks and gluons attached to each other.

Protons and neutrons situated inside a nucleus are constantly moving, continuously vibrating and oscillating, and this agitation significantly reduces the degree of cohesion of the nuclei, especially of those which are heaviest. The behaviour of nuclear matter, despite the enormous intensity of the forces involved, resembles in certain respects a high-density liquid. And, like all fluids, nuclear matter too proves to be very difficult to compress further.

Combine protons and neutrons and you will have the nuclei of the various elements; put into orbit around these nuclei clouds of electrons distributed in shells, the *orbitals*, and you will have the corresponding atoms, held together by the electromagnetic force. Atoms which share electrons in their outer shells come together to form molecules, made up of atoms held together by the electromagnetic force. All it takes is a small leakage of the force which holds atoms and molecules together for there to be weak electromagnetic reactions between them, like the van der Waals force.

It is often said that modern science has merely confirmed what had already been instinctively understood by the great Greek scholars: Leucippus, Democritus and Epicurus. Although the statement contains some truth, it's important to emphasize the essential differences between classical atomism and the elementary particles of contemporary physics.

Above all, among the components of our Standard Model, an important role is played by the particles which transport the forces, and which resolve in an astonishing manner the problem of the aggregation of 'atoms', which philosophers and scientists had been racking their brains over for millennia.

As far, then, as the properties and behaviour of elementary particles is concerned and, as we shall see, of the void too in which they are propagated, this is a whole other story, so bizarre and innovative that not even the most inventive minds among the great thinkers of antiquity could ever have imagined it.

And that is what we shall discover in the next chapters.

4

Clouds, soft matter and the last shamans

The Kunsthistorisches Museum in Vienna is home to many masterpieces. The great palace near the Hofburg, the Imperial Palace of the Habsburgs, hosts a collection of wonderful works by Raphael, Titian, Rembrandt, Vermeer and so on. When there is such an abundance of beauty, there is always the risk of neglecting equally admirable works, which are simply hung in more out-of-the-way corners. *Jupiter and Io*, an oil on canvas by Correggio, can be found in a long side corridor of the museum, away from the main thoroughfare of the great rooms, but no visitor walking past this painting can remain unaffected. Everyone stops, young and old, because the subject is so intriguing and is executed with such mastery.

As often happened in the sixteenth century – the best-known case being that of Michelangelo Merisi, known as Caravaggio – Antonio Allegri's contemporaries also referred to him by the name of the small town where he was born, Correggio, near Reggio Emilia. The painter had carved out a decent reputation for himself throughout the region based on the solutions he had come up with while painting the frescoes in the cupola of the Church of St John the Evangelist in Parma.

His *Ascension of Christ among the Apostles* provoked astonishment and admiration. The depiction of clouds was one of the major challenges for painters of the period and Correggio had resolved it with unprecedented mastery. Apostles and angels, which form a crown around the figure of Christ in triumph, pose lightly on folds of transparent clouds, shot through with blazes of light.

The success of the work won him another significant commission for the Cathedral in Parma, but most of all it opened the doors to Mantua to him. Duke Federico II invited him to work at the Gonzaga court, one of the most coveted residencies for artists of the period.

It was Federico's mother, the powerful Isabella d'Este, one of the most important female figures of the Renaissance, who transformed Mantua into a capital of art and beauty. She had summoned to court Raphael, Mantegna, Titian and Leonardo da Vinci, who in fact executed a wonderful portrait of the Marchioness. The drawing in charcoal, red ochre and yellow pastel, today in the Louvre, is one of the finest cartoons ever to have been created.

For Correggio, being called to Mantua meant making a great leap, joining the ranks of the most important painters. Federico II asked him to paint the loves of Jupiter, taking as his inspiration Ovid's most famous work, the *Metamorphoses*.

In the painting on display in Vienna, Correggio represents the love of the king of the Gods for Io. The myth tells of this stunningly beautiful priestess of Juno, with whom, as was his wont, Jupiter fell madly in love. To escape the control of his fearsome consort, the lord of Olympus made a fog descend into the grove where the young woman was and, protected by the mist, was able to possess her.

Correggio chose to interpret the fog of Ovid's poem as a transformation of Jupiter into a cloud and executed this incredible work which depicts the magical and impalpable intercourse between a beautiful virgin and a dark cloud. Jupiter is vaguely outlined: the face of the sweetest of lovers conjoining with his beloved, who welcomes him with half-closed lips, while his powerful right arm wraps around her waist. The abandonment of Io's white and sinuous body, portrayed from behind, to the cloud's embrace is total, complete. An ellipse links the young woman's left arm and right foot, which wrap around the most powerful of the Gods, who has become something soft, capable of making his way into everything.

Jupiter and Io is one of Correggio's last works, a masterpiece of beauty which reveals the Emilian artist's incredible mastery of the painterly medium. The erotic abandonment of the young priestess to her lover, both powerful and impalpable, will be destined to remain forever in the collective imagination.

States of matter

Why can we mortals not transform ourselves into a cloud? If we were suddenly able to become something rarefied and fine, infinite possibilities would open up. And not just in repeating Jupiter's exploits, slipping unseen into lovers' dwellings and then disappearing in the blink of an eye, should danger call. For certain it would be more complicated avoiding thefts, and the protection of treasures preserved in bank vaults would become a major problem. But if we did have this super-power, we could also use it for more noble aims: preventing us from getting hurt, for example, when tumbling downstairs or plummeting off scaffolding. Or to survive a car accident or a plane falling out of the sky.

Unfortunately, we aren't granted this capacity, because our bodies have a material consistency which affords us the possibility of brushing against a rose petal or caressing a baby's cheek, but not of walking through walls.

Like everything which surrounds us, we, too, are largely made up of void. The volume occupied by the quarks, gluons and electrons which make up our body is negligible. The X-rays in an X-ray scan cross our bodies from one part to another and the same is true of other less well-known forms of radiation, like cosmic rays. But around the nuclei are clouds of electrons which surround the atom and group together to form molecules. Whenever we touch any material object, the so-called contact is nothing other than a repulsion between two external electron shells, in which charges of the same sign are circulating and, as a result, repel each other. As if that were not enough, there are further rules – we shall see them later – which prevent electrons from occupying the same positions.

But then, one could imagine changing the structure, effecting a transformation of state such as to render our internal constitution less cumbersome. A fine and highly penetrating gas, for example the helium used to inflate children's balloons, can effortlessly move through hermetically sealed steel walls. Its minuscule molecules manage to slip through the tiniest gap left free in the crystalline structure. Unfortunately, even in this case we will discover that there are no short cuts for us either.

Let us try to understand the scenario better by starting with the different states of ordinary matter, conventionally referred to as solids, liquids and gases. For the moment, let's set aside material forms such as plasmas which, although they play a fundamental role in the structure of the universe, only occasionally feature in our everyday experience.

To fully understand how ordinary matter is organized, we need to make use of binding energy. This concept stems from a consideration according to which, when two material bodies are bound together, energy has to be used to free them from this condition. This is a general principle and is valid for systems which are very different from one another: for example, for the protons confined inside a nucleus, for the two hydrogen atoms bonded with oxygen in a water molecule or for a communications satellite in orbit around the Earth transmitting the World Cup Final.

In each of these cases, to free the two components of a system from the force of attraction which binds them together, energy needs to be expended. For example, if we wished to free the satellite in orbit around the Earth from the force of gravitational attraction which keeps it bound to our planet, we should have to equip it with a powerful rocket motor. By firing the latter, the satellite would increase its velocity and would be able to move away from its pre-arranged orbit. If it had sufficient fuel, it could reach its escape velocity, that is the velocity that would allow it to escape completely from Earth's attraction, and then travel further into deep space. For every bound system, one can calculate the energy necessary to make the two components of the system completely free of each other; this energy corresponds to the binding energy of the system.

The most convenient units for measuring the energy of microscopic systems are multiples of electronvolts, eV. One electronvolt is defined as the kinetic energy acquired by an electron which, from a static state, is accelerated in a vacuum by a potential difference of 1 Volt. We are dealing with a minute quantity of energy, for which reason it is convenient to use its multiples: 1 KeV, a thousand eV or kilo-electronvolt, that is 10^3 eV; 1 MeV, a million eV or mega-electronvolt, that is 10^6 eV; 1 GeV, a billion eV or giga-electronvolt, that is 10^9 eV; 1 TeV, one thousand billion eV or tera-electronvolt, that is 10^{12} eV.

Even in the case of the most impressive multiples, we are still talking about negligible energies compared with those which characterize the macroscopic world. For example, the protons that are circulating currently in the LHC (the Large Hadron Collider, the largest and most powerful particle accelerator), at CERN, have an energy of 6.8 TeV. They are the most energetic particles that we have managed to produce with the most advanced of our accelerators, but their energy corresponds to that of an annoying mosquito that one summer evening has chosen to bite us on the neck: barely perceptible.

By using multiples of eV, it's easy to describe the binding energies of different systems. For example, the energy that binds two atoms that form a molecule has typical values of a few eV. The energy that holds back electrons and forces them to orbit the nucleus varies from a few eV, for electrons most weakly linked to external shells, to around 10 KeV for those in the innermost shells.

When we move from electromagnetic bonds to those produced by the strong force, there is a sudden jump. The binding energy of a single proton or neutron in a nucleus is around one million times higher than that which holds onto one of the outer electrons. The energy between protons and neutrons of an atomic nucleus is measured in MeV, whilst 1 GeV, that is energy which is a thousand times greater, is needed to shatter a proton and free some of the quarks and gluons of which it is composed.

A bound state always corresponds to a total energy lower than that of its free components because part of the system's energy is used to hold them together. Since $E = mc^2$, the equivalence between energy and mass discovered by Einstein, applies, this difference can be found as a difference in mass. Mass and energy, of equivalent size, can both be measured in eV, by means of its multiples.

Thus, if we take a free proton and neutron, their total mass is greater than that of the deuterium nucleus, a kind of heavy hydrogen, which has a nucleus formed of one proton and one neutron bound together by the strong force. The difference in mass is around 2 MeV, which is precisely the energy of the proton–neutron link. If we wish to break apart a deuterium nucleus and go back to having a free neutron and electron, we need to supply them with little more than 2 MeV of energy. The binding energy of a system is always negative, which is the property of the bound state; to demolish the prison cell in which the proton and neutron are enclosed, we have to use energy.

When systems are held together by very small binding energies, this difference in mass becomes barely significant. This is the case for chemical bonds, those established between atoms to form molecules or crystalline structures, or between molecules that interact. These are due to electromagnetic-type forces, and the relative binding energies can vary from several tens of eV right down to a tiny fraction of an electronvolt. When the bonds are so weak, they can easily be broken even by just heating the system, thus supplying energy in the form of heat.

The heat that two bodies exchange is connected to the kinetic energy of the atoms and molecules that make them up. Depending on the case, these constituents can move freely or vibrate and oscillate around their equilibrium positions. Placed into contact with a warmer body, whose components have a higher average kinetic energy, they receive energy in the form of mechanical impacts, with the result that the cooler body warms up and the warmer one cools down, until they find a new equilibrium. Through heating, energy can be ceded to systems of molecules or atoms to the extent of breaking the internal bonds between them. In these conditions we are present at what is known as state transitions. The same substance, without changing its chemical nature, assumes a different internal construction.

On the basis of this mechanism, we can distinguish between the three traditional states of matter: solid, liquid and airy or gaseous. The ice cube in the glass of orange juice, the saucepan of water for the pasta that begins to boil, the leftover meat broth that is put in the freezer to be used later, maybe to make a delicious risotto, are all experiences of a change of state that are part of our daily lives.

Water is an example, familiar to everyone, of a substance that can present in three completely different states, depending on the conditions of pressure and temperature. To simplify things, let's suppose that the pressure always remains constant and let's recall that for a determined state of the body, its temperature is linked to the average energy of its components: atoms and molecules.

By heating an ice cube, the bonds between the water molecules that produced the frozen crystal, are broken and the molecules slip over each other, interacting weakly with those around them. In the liquid state, the water becomes a fluid, capable of changing shape and adapting to any receptacle. As the temperature rises, when the remaining bonds between molecules are broken, these can move freely, colliding with each other now and again, in the entire volume occupied by the steam, which has increased massively. By inverting the flow of heat, that is to say by reducing the heat in the water vapour, it is possible to produce transformations in reverse, making it initially condense into liquid form, which can then be frozen to form a solid block of ice.

But the water vapour is exactly the cloud into which Jupiter transforms himself to give vent to his passion for Io. Why can't we do the same? Why can't we go into a very hot sauna and come out in the form of impalpable steam?

The strange world of soft materials

The matter our bodies are made of belongs to a category of material systems that is not easily ascribable to the three conventional categories of solids, liquids and gases. Like all biological systems, our body is a very complex and highly composite organism. It is formed around a robust scaffolding, the bones of the skeleton, which supports a heterogeneous ensemble of soft tissues, imbued with liquids and with a high water content, whose functions are managed by the central nervous system.

The soft biological material of which we are composed offers many advantages. We need only think of our hands and the tips of our fingers, thanks to which we can caress a new-born baby's face, lightly touch the keys of a piano, model a statue out of clay or appreciate the smooth surface of porcelain. The skilful use of hands is the secret behind the most sophisticated artistic works, the most moving musical performances, the most precious miniatures, the most refined jewellery, the most exquisite paintings. Having bodies made of soft materials, elastic joints and flexible material structures allows us to run or dance; it makes embracing each other or cuddling our little ones pleasurable. As a result of this characteristic, we have developed an extreme sensitivity to touch. Our fingertips are capable of sensing the least roughness on a surface or of recognizing a vast quantity of materials, even without the aid of sight or smell.

The use of hands has played a fundamental role in human evolution ever since the moment, several million years ago, when one of our distant forebears adopted an erect posture. Once we were bipeds, our front limbs, finally released from the constraints of locomotion, assumed functions that were more and more important. Our hands became the instrument for gathering food and putting it in our mouths, for making ever more complex tools; in the same way, our mouths, freed from the inconvenience of grabbing, tearing up and gathering food, ended up concentrating on another activity, equally fundamental for the survival of the social group: articulating sounds and constructing a language.

This is a crucial turning point in our evolution, which set in motion other decisive events. The sensitivity and versatility of our hands, together with the development of language, kick-started our neuropsychic development. The brain not only increased in size, its internal structure and its functionality became ever more sophisticated. It's no coincidence that still today, the part of the brain that is concerned with controlling fingers, hands, lips and tongue, is around half the total motor cortex, the part which presides over all our movements.

With our hands, we have managed to produce ever more versatile tools, which have increased our chances of survival. For thousands of generations, we have used our hands to caress each other, console each other, to tend the weakest, to seek mutual comfort, to scratch ourselves or remove each other's fleas. The hand, originally a kind of rudimentary pincer for grabbing stones, has become an instrument of social cohesion and at the same time emblematic of the creative capacity of the human imagination: all fundamental activities for building society and strengthening the group.

It's no coincidence that hands often feature among the earliest subjects of artistic representation. Hundreds of coloured handprints decorate the walls of caves inhabited by our ancestors. They are formed as outlines, in the negative, with pigments blown or spat out by the youngest members of the group, often girls, who, in so doing, left something of themselves on the walls – an inanimate sign but one destined to last, a testimony to their existence.

Fingertips, with their characteristic patterns of fine grooves – our fingerprints – vary so much from one human being to another as to become a kind of personal identity document. The underlying skin is so extensively dotted with nerve endings as to represent a kind of immensely powerful universal sensor: a sophisticated and wonderful structure that can compete, in terms of its powers, with miracles achieved by nature in other species, such as in the case of geckos' toes. The feet of these small reptiles, covered with millions of tiny hairs, are capable of exploiting the weak van der Waals force of attraction between molecules; this tiny animal can, thus, move with agility on any vertical wall and even remain attached to the ceiling with its head facing down.

Unfortunately, however, the soft materials which compose the bodies of living creatures, however versatile and efficient, are intrinsically fragile. It takes barely anything to produce wounds and lacerations and they are extremely sensitive to heat. If we immerse our fingers in boiling water, or worse still come into contact with a flame, we suffer very serious burns. All of this has to do with the general properties of soft materials, whose behaviour is in some ways intermediate, somewhere between that of a liquid and that of a solid.

To understand how they function, we can think of some very common soft substances, such as butter, mayonnaise or toothpaste. These are all materials that can easily be reshaped or even spread. They are mixtures of two or more heterogeneous components that are not soluble with each other. In the case of butter, these are animal fat and water; mayonnaise, on the other hand, is obtained by blending the yoke of an egg with oil and lemon juice, while at the same time incorporating a large number of air bubbles which soften it.

All these materials are very sensitive to heat. Subjected to excessive heat, they change state irreversibly, burning and falling apart. This is what happens to an egg when we put it in the frying pan. At room temperature, the white of the egg presents as a slimy, transparent liquid, made up of water in which proteins are dispersed. When we turn on the gas, the heat causes the proteins to bond with each other, thereby forming an opaque, elastic mesh, typically white in colour. All these transformations are irreversible; nobody has ever succeeded in returning a fried egg to white and yolk by putting it in the fridge.

This characteristic, which makes changes of state for biological tissues irreversible, is connected to the complexity of living matter. The brilliant organization, which allows matter to come into contact with the surrounding world, to extract nutrients to help it grow, reproduce and colonize new environments, turns out to be so complex that it is inevitably destined to break down in time. The death of every living species is nothing more than an ineluctable process of oxidization and simplification of the biological structures.

This same process does not affect inanimate matter or, at least, this is the assumption that we, anthropomorphic apes, have held for a very long time. We have been living on Earth for too short a time, and not one of us, or even one of our most distant ancestors, has ever witnessed a cosmic catastrophe capable of destroying an entire planet or a large star like the Sun. Nobody has seen from close up a dying star expand to the point of absorbing the planets orbiting around it and vaporizing them in an instant.

It's only a few years ago that astronomers and astrophysicists constructed instruments capable of showing us the dark side of the cosmos. Since then, we have been able to observe enormous black holes swallowing up entire solar systems, neutron stars colliding with each other, relativistic jets of matter which shatter whole galaxies. But too little time has elapsed. This knowledge has not become common wisdom, and the assumption that the great material structures of the cosmos will endure, an assumption that has gone hand in hand with humanity for millennia, remains very widespread.

The almost eternal life of the great material structures

The life of a planet, a solar system, a galaxy is measured in billions of years. This timescale is so disproportionate compared to those familiar to us that we really struggle to conceive of it. The secret of such vastly long existences is concealed in the structure of the atomic nuclei of the materials of which they are composed and in the evolution of the temperature of the universe over time.

The hydrogen and helium nuclei which form the stars and large gaseous planets like Jupiter, or the nuclei of the heavier elements up to uranium, which compose rocky planets like Earth, are made up of protons and neutrons, and their incredible resilience derives from the robust structure of these minute particles.

They are both composed of up and down quarks held together by gluons, but with one small difference; the proton is charged whereas the neutron is neutral, and the mass of the proton is slightly smaller than that of the neutron. This property has significant consequences for the stability of the great material structures; protons cannot decay into neutrons and electrons because this decay would violate the conservation of energy. Other examples of decay, possible in theory, have never been recorded. And so, as far as we know today, protons are condemned to live for eternity, or at the very least for a time that would exceed the current age of the universe by many orders of magnitude.

The protons that formed in the first minutes after the Big Bang were able to drift freely for hundreds of millions of years before joining together to create the first stars, which in their turn, billions of years later, grouped together into the first galaxies. That immense primordial population of protons is still alive today, here all around us, feeding the myriad of stars which light up the sky and building the atomic nuclei that we and everything that surrounds us are made up of.

On the other hand, neutrons, if left to themselves, that is isolated from the nucleus, would almost all die within the space of a few hours. They are spared this sad fate by the fact that they are bonded to protons to form the atomic nuclei of the various elements. In the comfortable shell of the nucleus, busy exchanging residual strong force with the other neutrons and the protons, and focused as they are on remaining glued to everything else, they cannot decay.

The structure of the three quarks held together by the strong force is particularly robust. To understand this, we need to make some observations about the masses of the particles. For example, the extremely light electron weighs around 0.5 MeV, whereas protons and neutrons have masses slightly lower than 1 GeV and are thus around two thousand times heavier. The two lightest quarks are heavier than the electrons, but much less than the protons; the up quark has a mass of around 2 MeV and the down quark of around 5 MeV. And so, we can immediately understand that the entire mass of the proton comes from the tremendous energy produced by the gluons which interact among the quarks. Around 1 GeV of binding energy means that there is an incredibly strong glue which holds the components together. This little structure is very well put together; no expense was spared in making it solid and capable of resisting the most powerful disturbances.

To break a structure that is held together so well, you need highly energetic particles or truly monstrous temperatures. Such phenomena can be found in the cosmos; there are mechanisms capable of accelerating particles to energies even many orders of magnitude higher than required, but they are rare. The *natural* movement of these high-temperature particles, which are capable of smashing protons and neutrons, is fairly limited, and thus the matter inside stars and large celestial bodies can serenely survive the passing of time.

Heat – vast heat – could split apart the magic box inside which the quarks are hiding. But they would have to be heated to temperatures of a thousand billion degrees. It would be impossible to do this even for the most massive stars, whose internal temperatures never exceed several hundreds of million degrees.

As the universe evolved, it experienced terrifying temperatures, in particular during the earliest moments of its life. But it cooled down very rapidly, while expanding at great velocity. Just a minute and a half after the Big Bang, we had already dropped below ten billion degrees, a comfortable temperature for the components of the nuclei which from that moment onwards were able to survive without difficulty and even begin to join together. Today the average temperature of the extremely old universe in which we live doesn't get as high as three degrees above absolute zero, that is to say approximately minus 270 degrees centigrade, and there is no danger that the protons will fall apart.

This characteristic of great celestial bodies to survive over the ages and to remain unaffected by ordinary variations of temperature and pressure, lies behind the assumption that the great material structures will last forever.

Since remotest times, humanity has experienced its own intrinsic fragility as something to be ashamed of. Our being mortal, a characteristic we share with all living beings, has led us to idealize all enduring material forms: oceans and mountains, rivers and volcanoes, as well as the Moon, the Sun and the stars. Over the centuries we have evolved an assumption that inanimate matter survives for millennia and defies mortality.

This concept was thrown into crisis at the beginning of the twentieth century, when physicists began getting to grips with the most ephemeral forms of inanimate matter. They discovered a world of material states that were so precarious as to make the existence of even the most unfortunate of insects, the ephemerides, almost enviable, their very name implying the insignificant duration of their life. These tiny aquatic insects, similar to dragonflies, live for around a year in the larval state, but once they reach adulthood, the rush is on to find a partner to mate with, because they have just a few hours of life remaining.

The evanescent world of the most ephemeral forms of matter

The existence of these ephemerides is infinitely longer and more varied than that of many states of inanimate matter. The first suspicions that even ordinary matter could decompose spontaneously emerged at the end of the nineteenth century, when the French physicist Antoine Henri Becquerel and the famous Polish/French scientist couple of Maria Skłodowska and Pierre Curie, began studying natural radioactivity.

They noted that some very heavy elements, that is to say those made up of nuclei containing a large number of protons and neutrons, were unstable and emitted various types of radiation, decaying and transforming into other elements. This was the beginning of the long road which led to a deeper understanding of the structure of matter.

Unstable elements became ever more numerous when it was discovered that radioactivity could be induced by bombarding certain elements with neutrons. Other unstable particles were identified in cosmic rays and above all dozens of them were discovered when the first particle accelerators started working.

Little by little as the energy of the new machines grew, it became possible to exploit the transformation of energy into mass to produce new states of matter. And new particles were discovered that were very different from the ones that make up ordinary matter; their behaviour was rather strange and, most importantly, they decayed in much shorter times. From the 1950s onwards, the catalogue of the components of matter was enriched with an impressive quantity of new particles; these were almost all highly unstable, and some had an existence so brief as to make it impossible to measure their duration directly.

The secret of this world of ghost particles, which appeared for a fraction of a second at the centre of experimental apparatus, merely to disintegrate, immediately exploding like a mini firework, lies in their composition. In the 1960s and 1970s, physicists realized that the ephemeral matter being produced in their accelerators, despite playing a marginal role in the world of ordinary matter, was fundamental to understanding its origins and to reconstructing the complicated evolution that had carried it down to us. Used to scampering around freely in an incandescent, new-born universe, these massive particles didn't survive even for an instant in the ultra-freezing environment in which they found themselves, but that instant was enough to study their properties.

To understand in depth the laws that govern ordinary matter and the material composition of the universe as a whole, physicists have had to travel through the world of ephemeral matter and reconstruct all its properties. Like ancient shamans, who acquired knowledge by penetrating the world of illusions and dreams, modern scientists too have discovered the most profound symmetries in nature, by venturing into the phantasmal world of the most evanescent and fleeting material states.

The Standard Model of elementary particles extends to the exotic states of short-lived matter the same approach that allowed us to explain ordinary matter. The generalization of the model includes three generations of quarks. The first of these is the one we already know, formed by two quarks, up and down, which form protons and neutrons; the other two contain much heavier quarks with rather peculiar names.

The second generation is made up of the *strange quarks* (with a charge of −1/3) and *charm quarks* (with a charge of +2/3) both of which are very heavy. The charm quarks in fact weigh more than a proton with a mass hundreds of times greater than that of light quarks.

The third generation contains the champions of their category: *beauty (or bottom) quarks* (with a charge of −1/3) and *top quarks* (with a charge of +2/3). The top quark has the heaviest weight in the category: it weighs more than 170 GeV, as much as a gold atom. Like the up and down quarks, the quarks of the other two generations are imbued with an electric charge, a weak charge and a colour charge.

All quarks can combine with each other in various combinations, giving life to a complex zoology of hundreds of exotic states of matter held together by the strong force. The particles that derive from this are called *hadrons*, from the Greek αδρός (*hadròs*), strong, to indicate that they are held together by the strong force. Depending on the number of constituents, hadrons divide into *mesons*, when formed by a quark-antiquark pair, and *baryons*, when they are made up of three or more quarks. Neutrons and protons are the most common baryons and the only stable ones. Matter that is formed when heavy quarks join together is highly unstable and decays rapidly into other particles containing only the lightest quarks.

For every particle in the Standard Model, we always need to consider its corresponding anti-matter partner, a particle which has the same mass and several opposing quantum numbers, including the electrical charge. For example, the *anti-up quark* has the same mass of around 2 MeV as the up quark, but a charge of −2/3.

The first to predict anti-matter was the English physicist Paul Adrien Maurice Dirac. In 1928, among the solutions to an equation that was to enter the history books, he saw one emerge which seemed to describe the motion of a positive electron, a particle very similar to the one which orbits around the nuclei, but with an opposite charge. Initially this didn't spark too much astonishment; most people considered it to be a mathematical curiosity, until a young American physicist noticed something very strange in his experiments on cosmic rays. In 1932, while trying to understand the composition of the mysterious flow of particles coming from the deepest part of the cosmos, Carl David Anderson, twenty-seven at the time, was baffled; among hundreds of known particles, he found one which had all the properties of an electron apart from its positive charge. He named it the *positron*, without knowing that this discovery would earn him the most prestigious of scientific accolades. In 1936, when he received the Nobel Prize, Anderson was still only thirty-one, becoming one of the youngest laureates of all time, only beaten by the Australian-born British physicist, William Lawrence Bragg, who had been awarded the prize in 1915 at the age of twenty-five.

Since then, further discoveries of *antiparticles* followed regularly, and many of their properties were confirmed. It was observed that, when an electron encounters a positron, the phenomenon of annihilation occurs, with the two particles disappearing and making way for two photons. The existence of the opposite phenomenon was also ascertained: that is that high energy photons can produce pairs of electrons and positrons and the same can occur with gluons capable of creating pairs of quarks and *antiquarks* 'out of nothing'.

These processes lie behind modern accelerators, known as *colliders*, which by means of the annihilation of elementary components succeed in producing new, extremely massive, particles. For example, LEP, the Large Electron-Positron Collider, which was operating at CERN before the LHC, made electrons and positrons collide at high energy to produce Z, the neutral particle which carries the weak interaction, and which has a mass greater than 90 GeV.

The electrons that orbit around the nucleus are part of the *leptons*, from the Greek λεπτός (*leptòs*), meaning thin, light, because they have small mass. Leptons, too, are subdivided into three generations, each formed by two particles, like quarks, but in this case one is charged and the other neutral. The first generation is formed by the electron and the electron-neutrino. The latter is a neutral particle, so light that until some time ago it was thought to have zero mass.

Like the electron, the neutrino that accompanies it also has no strong charge and, as far as we know, is a stable particle. Since it is electrically neutral and very light, its exchange of forces with matter is limited to the weak interaction and gravitational effects, which are insignificant since its mass is tiny. Neutrinos can travel through the whole of the Earth from side to side leaving no trace of their passage. It's hardly surprising that such delicate, fine particles should have escaped our attention until a few decades ago.

The second lepton generation is formed by the charged *muon* and its neutral muon-neutrino. The muon is a kind of very heavy electron: its mass is two hundred times greater, measuring around 100 MeV.

The third generation is formed by the *tau* lepton, the true champion in its category because it weighs almost as much as two protons, and by its neutral partner, the tau-neutrino. Muons and tau leptons are not stable; the weak force makes them disintegrate rapidly into lighter particles.

Quarks and leptons are rather alien families which do not willingly mingle. Acting as a bridge between the two constituents of matter are mediators or force carriers, a third family whose components interact with both groups, sometimes only with some of the various members, producing dynamics and mingling.

We have already spoken about mediators: they are the photon, the carrier of the electromagnetic force which acts on all particles with an electric charge, the gluon, which transmits the strong force and interacts with quarks which have a strong charge but ignores the leptons which have no charge, and finally W and Z, carriers of the weak force, which combine with either quarks or leptons, because all of these have a weak charge.

As opposed to photons and leptons, which have zero mass, W and Z are extremely massive. They weigh respectively around 80 and 90 GeV, and this, as we have seen, limits the radius of action of the weak force to sub-nuclear distances. This last interaction does not lead to bonded states, but is responsible, as we shall see, for many processes fundamental to the evolution of the universe.

There is, finally, one other component of the Standard Model which stands slightly on the side-lines. It is the most recent arrival and does not belong to any of the families described thus far: this is the *Higgs boson*, which plays a decisive role in the constitution of matter as we know it. We shall talk about this at length in the next chapter.

5

The triumph and decline of a thousand-year-old assumption

More than ten years have passed, but it still doesn't seem real. I often find myself waking up in the morning and wondering whether what happened to us really did happen.

The discovery of the Higgs boson was announced to the world in July 2012, when we shared the news that, at CERN, we had found a new particle, which resembled in many ways what had been nicknamed the Holy Grail of physics. Thereafter, by gathering further data, it was possible to put together a better reconstruction of the identikit of the newcomer and the final doubts faded away. Yes, this really was the one. The Royal Academy of Sciences in Stockholm was also convinced and, in 2013, awarded the Nobel Prize to Peter Higgs and François Englert, the two physicists who had hypothesized its existence almost fifty years before.

The life of an experimental physicist is strewn with challenges and great risks. Doing research at the very edge of knowledge means heading down new paths that have never been trodden before, and accepting the possibility of failure. In modern science, nobody embarks on a new adventure without taking potential failure into account. We particle physicists are passionate and curious people, but in our environment only those who love challenges survive, only those who have no fear of rolling with the punches and finding themselves on the floor. When that happens, and it happens very often, there's no time for whingeing. You just have to pick yourself up, understand where you went wrong and prepare to start all over again. And so, one of the most prized qualities in our field is resilience.

Ours is a high-risk profession; when you start out on a new piece of research, not only are you not sure of achieving the predicted results, you are not even sure you'll manage to get things off the ground, because you have to choose the technology to use, and maybe it won't work. You also have to raise the funds and maybe you won't succeed. And finally, those ideas, which at the outset seemed so brilliant, might turn out to be completely wrong.

None of us does this work for honours and recognition, even if we're delighted when they come our way. The true satisfaction is being there, in front of our computer monitors, aware of being among the first human beings to contemplate a new state of matter. The emotion you feel at moments like that is difficult to describe, but it is its own reward for all the sacrifices, fears and dangers that you've had to face.

I belong to a very fortunate generation of physicists, fortunate because it doesn't often happen that you find yourself in the front line of those contributing to such an important discovery. You can spend your whole life chasing a result without ever achieving it. And that doesn't mean that you're not good enough.

Let's take the generation of scientists who preceded us, those who were seeking the Higgs boson for decades, starting in the 1970s. They weren't successful, not because they weren't as good as we were, but quite simply because the conditions weren't in place. The accelerators of the period were just much less powerful than the LHC. Today, now that we have discovered it, we know that the Higgs boson is a very heavy particle. Managing to produce and identify it would have been impossible with the machines that were in use back then.

We just happened to be lucky enough to find ourselves in the right place at the right time. Unquestionably the effort made by thousands of scientists to build the LHC and its large detectors was extraordinary. The young especially – the physicists, engineers and computer scientists who make up the assault force of these collaborations – came up with so many innovative ideas and were thus able to overcome the enormous difficulties encountered during the construction and the roll-out of these technological jewels. But everything in our field is so complicated that neglecting one small, apparently insignificant detail would have been enough for success to have turned into colossal failure. And so, fortune, as always in human affairs, played a decisive role in this case too.

But what is a boson, and why is this discovery so important? And above all what role does this famous particle play in the organization of matter and the constitution of our universe?

Stories of fermions and bosons

Through various combinations, the particles of the Standard Model can form hundreds of different material states. It's a little like assembling complex objects using Lego bricks which, though very versatile, can't be joined together randomly; rules have to be followed.

The same is true for interactions between quarks and leptons. Here too there are rules to be followed, quite simple rules, which are referred to as the 'conservation of quantum numbers'. For example, you cannot combine three quarks and end up with a state that has fractional charge. The charge must always be whole and must be conserved in the reaction processes between particles.

The two families of matter particles, quarks and leptons, largely reluctant to mingle with each other, nonetheless have a property which draws them together; they have fractional *spin*, in this particular case of value +1/2 or −1/2. Spin is another quantum number which is conserved in the interaction between particles, one of their characteristic properties for which no classical equivalent exists. It can be interpreted as a form of intrinsic angular momentum, as if the matter which makes up the particle were behaving in the same way as a tiny spinning top turning around itself, but the analogy is rather imprecise. Electrons, which are practically punctiform, have half-integer spin; it's difficult to imagine rotating something around an axis which is wholly concentrated into one point, therefore lacking any dimensions.

Mediators – the particles that carry forces – on the other hand, have integer spin (0, 1, 2), for example photons, gluons, W and Z have a spin of 1. Don't ask me the reason for this difference. Nobody knows why. Some speculate that there is an underlying, deeper symmetry that we haven't yet discovered. Maybe there is another form of matter in which the roles are reversed: super-matter in which the super-particles of matter have integer spin and those which carry the interactions have fractional spin. That is a material world which, on paper, could work, but nobody has yet come across even one of these super-particles. For now, we have to accept this division, which turns out to bring with it many important consequences, because having fractional or integer spin leads to fundamentally different behaviour and interactions.

Particles that have fractional spin are called *fermions*, whereas those that have integer spin are called *bosons*. There are composite fermions, which have a spin of 3/2 or 5/2, and bosons, which have a spin of 0 or 1; there is a hypothesis that elementary bosons with a spin of 2 also exist, but for now these haven't been discovered.

The strange names of the two families derive from the laws of statistics which the two types of particles obey, and here there is a decisive difference. Fermions follow the rules defined by Enrico Fermi and Paul Adrien Maurice Dirac, whereas bosons follow those theorized by the Indian physicist Satyendranath Bose and Albert Einstein. The difference relates to Pauli's principle of exclusion, named after the physicist Wolfgang Pauli, who formulated it in 1925. This is a norm which is not valid for bosons, whereas it turns out to be a kind of inviolable tabu for fermions; two identical fermions cannot occupy the same quantum state. With a little imagination, it can be conceived of as a kind of agoraphobia; quarks and leptons cannot tolerate finding themselves in a quantum state which is too crowded. The crowd sends them into a panic, and they have to rush off to find a less congested environment. Put like this, it would seem rather exotic, lacking in any practical consequences. The reality, however, is quite different.

Because of the exclusion principle, electrons are forced to occupy different atomic levels, and this allows for the formation of molecules and chemical reactions. The orbital closest to the nucleus, which corresponds to a clearly defined quantum state, cannot be occupied by more than two electrons: one with an up spin of +1/2 and the other with a down spin of −1/2. Other electrons have to occupy higher orbitals and as these too gradually become filled, they are forced to occupy orbits further and further from the centre, where the binding energy with the nucleus diminishes, and it takes very little for them to be ripped away or forced to exchange links with other atoms. Thanks to this mechanism, atoms can interact among themselves. Without Pauli's exclusion principle, all the electrons would only occupy the innermost orbital and each element would behave like a kind of noble gas, inert and incapable of interacting with anyone. In the world there would be no more molecules, nor even physicists trying to understand how matter functions.

None of that is valid for bosons, which can crowd out the same quantum space without any limits. As a result of this property, among other things, we have been able to develop lasers, beams of coherent light in which an unlimited number of photons share the same quantum state. It could be said that, if we hadn't understood the laws that govern the life of bosons, we would not, today, be able to cure myopia, read the labels on supermarket products or transmit information via fibre optics. As we shall see later, Pauli's exclusion principle plays a really important role in the evolution of the stars, too, and the fact that bosons are exempt from it has even allowed us to build new states of matter.

But what really is mass?

One of the cornerstones of the Standard Model is the unification of electromagnetic and weak interactions. The theory is built around the hypothesis that these two forces, apparently so different from each other, are in fact manifestations of the same force, the *electroweak* interaction. In the twentieth century, a hundred years after the unification of electricity and magnetism, history repeats itself.

Everything stems from a formal analogy which reinforces the intuition from which Fermi started. The equations that describe the two forces are similar to each other and this correspondence cannot be coincidental. It was the American physicists Sheldon Glashow and Steven Weinberg and the Pakistani Abdus Salam who were the first to speak of an electroweak force and to construct a theory which hypothesized as mediators, alongside the photon, a trio of other bosons, two with a charge, called W+ and W-, and one neutral one, Z. Several years later, Gerard 't Hooft, a young Dutch student who was working on his thesis, managed to demonstrate that the theory was renormalizable, that is to say that it didn't run up against those contradictions that are a nightmare for any theorist. At that point, final doubts were overcome, and everyone ended up convincing themselves that they were on the right path. Also because, in the meantime, experimental tests confirming the predictions of the Standard Model multiplied, including the discovery of W and Z by Carlo Rubbia at CERN in 1983.

The success of this theory led to Nobel Prizes being awarded to all of the protagonists of this great adventure: Glashow, Weinberg and Salam in 1979, Rubbia in 1984 and 't Hooft in 1999. From then on, the predictions of the Standard Model have been subjected to ever more stringent tests and verified with impressive precision.

The theory, however, concealed a problem that would remain unsolved for decades: the question of mass. We have already seen that the extremely light photon makes the radius of action of the electromagnetic force infinite, whereas the massive W and Z are incapable of propagating the weak force beyond sub-nuclear distances. But if the Standard Model is based on the assumption of electroweak unification, how can it happen that the photon, with no mass, can carry the same interaction as W and Z which weigh as much as a hundred or so hydrogen atoms? What mechanism gives mass to W and Z and differentiates them from photons? In the final analysis, what exactly is mass?

In the mid-twentieth century, scientists found themselves confronting a question that had escaped even the greatest scientists' attention for centuries. A question that concerns such a well-known physical quantity that nobody had ever asked themself the question whether, beneath its apparent ordinariness, something less obvious was hiding.

The measurement of mass is perhaps the only physical measurement carried out by the whole of humanity for thousands of years. You don't have to be a scientist to stand on the scales and check your own weight or go to the market to buy a kilo of apples. The mass of a body is taken for granted to such an extent that a very persistent assumption has developed around it.

The origin of the word 'mass', which derives from the Greek μαζα (*maza*), indicates a mixture of cereals. Any material object has mass, for which reason it has always been considered an intrinsic property of bodies. Nobody ever asked themself if it was an acquired property, rather, that is to say, than a primal one.

This issue had eluded not only philosophers and great thinkers, but also the giants among modern scientists, who had concerned themselves with mass in various contexts. Like Galileo, who understood that the acceleration of bodies in free fall does not depend on their mass. Or Newton, who formulated the law of universal gravitation and understood that mass is the source of the force of attraction between two bodies. With Newton, mass loses that connotation of being an inert substrate; it becomes something living, interacting with the whole universe, but still remains an intrinsic property of matter. The assumption holds even with Einstein, one of the most imaginative minds of all time. Equivalent to energy, with Einstein mass becomes a much more concentrated form of energy, which grows with the velocity of bodies. No longer a static attribute, but variable, continuous and intimately linked to the dynamics of material bodies. In fact, in general relativity, mass-energy becomes an active material substance, which curves space-time and interacts dynamically with every other concentration of mass-energy wherever it's distributed. But not even Einstein is teased by the suspicion that beneath that concept, which was so familiar, there might be hiding something much deeper, an understanding of which would allow us to better comprehend how our universe was formed.

To supply solid foundations to the Standard Model of elementary particles, we needed to find the mechanism that had broken the electroweak symmetry, which had, that is, left the photon with zero mass and made W and Z massive. It took three young physicists to bring crashing down the centuries-old assumption that mass is an intrinsic property of matter and demonstrate instead that it emerges from an interaction with a new field. They were barely thirty when, in 1964, they proposed the theory which would forever change humanity's point of view about the concept of mass.

The generation of '64

In 1959, François Englert is twenty-six years old. It's the first time the young Belgian physicist makes it to the United States. He's there to work at Cornell University on a research contract. He's just touched down at the airport in Ithaca, the small city which hosts that prestigious academic institution, and there waiting for him is the young professor Robert Brout – thirty-one, a brilliant American theoretical physicist – whose assistant he will be. Of Jewish origin like François, Brout studied at Columbia, New York, and has only recently been taken on by Cornell. He goes to the airport to pick François up in an ancient Buick which was more rust bucket than car. They stop at a café to chat for a while. The conversation takes off at some pace. Their conversation continues throughout the night, at Brout's house, alternating coffees and spirits and discussing physics, girls, life and politics. When they leave each other to go and get some rest, they both have a sense that their friendship will last their whole life.

In 1961 Englert is offered a professorial chair at the Free University of Brussels and so returns to Europe. Sometime later, Robert transfers to the same university. He finds himself so much at home in Europe that he will end up renouncing his American citizenship to become Belgian.

The two physicists have lively and extrovert personalities. Endowed with a sharp sense of irony, they love good food and beautiful women, and they entertain themselves by playing tricks on their colleagues. They convey their passion to their students who are drawn in by their contagious enthusiasm. Their research interests encompass a number of areas, but in 1964 they decide to focus on what is going on in particle physics. It's a new field for them, as previously they had concentrated mainly on solid-state physics. They don't feel like specialists, and it is with some trepidation that they submit their debut article in this discipline to a major review in the sector. They're afraid they might have written nonsense or overlooked something essential. Because it seems to them that the question which the best theoretical physicists in the world are tormenting themselves over has an obvious solution, so simple as to seem self-evident.

They have seen a similar mechanism in action in other situations in solid-state physics, their field of expertise, and they suggest it also as a solution to the problem of the unification of fundamental interactions.

If the equations of the two interactions, electromagnetic and weak, are the same, what is breaking the symmetry between the carriers, the photon and W and Z, can be nothing other than the medium through which they propagate. The explanation is simple; the void is not a void. A new particle, endowed with mass, occupies every corner with its field.

It's easy to understand why nobody, in the first instance, took them seriously. Their article is published but passes unnoticed. This should be no surprise. The greatest experts in theoretical physics in the world are racking their brains to find a solution and these two novices, two unknowns, turn up with their hare-brained theory, claiming to be right.

According to their claims, every nook and cranny of the universe is full of something subtle and mysterious which only they can see and which only they have understood. The lack of reaction shouldn't astonish us. But the disappointment of the two young scientists is so great that, for some time, they consider turning their back on particle physics to focus on something else.

Chance would have it that, a few weeks later, the same journal receives another article, also written by an unknown young scientist. The work tackles the same theme from a completely different perspective, even if the conclusions are very similar to those of Brout and Englert.

The author, Peter Higgs, is a thirty-five-year-old English physicist, the same age as the Belgians but very different from them. He's a mathematical physicist in the early stages of his career, recently appointed as research assistant in Edinburgh. He is a shy, reserved type, almost misanthropic, and works alone; his colleagues consider him surly and slightly eccentric. His only passions are mountain-walking which he does with his wife, an American linguist with whom he is madly in love, and political activism.

He doesn't hide his ideas; from his time as a student in Bristol he has always been very active on the left-wing of the English Labour Party. He has supported union struggles and participated in CND, the Campaign for Nuclear Disarmament, wearing out the soles of his shoes on peace marches. He doesn't have a doctorate in particle physics, as a result of which he will always say, with a touch of snobbery, that he had no competence in the matter.

In the summer of '64 he writes one of his extremely rare articles, which he submits to a journal just before leaving for his holidays. In its first version the article is rejected and Higgs, reluctantly, has to set about working for a couple of weeks to adapt the script to meet the demands of the referee so as to make it livelier and more precise.

In the end his conclusions are clearer; yes, the spontaneous breaking of the electroweak symmetry occurs as a consequence of a scalar field which is produced by a new boson endowed with mass. And so, the article is accepted and appears in the journal a few weeks after the one written by Brout and Englert, which Higgs actually cites in his own work.

Many years later, in Stockholm, while we're toasting the medal which he had just been awarded, Peter will confide in me: 'I can't help thinking how strange the world is; if they hadn't rejected my article in 1964, I wouldn't be here this evening.'

The mechanism proposed to explain the separation of the two interactions is simple. When you see it summarized in a small number of formulae it appears almost self-evident. It seems like one of those brilliant solutions that nobody had thought of before, simply because it seemed too basic. Mass, the most common of the properties, conceals a trap. The ultra-light leptons and the heaviest quarks are all born, democratically, with zero mass. It's the Higgs' field which, permeating the whole universe, picks out and separates the massive particles from the light ones. The stronger the interaction with the field, the greater the mass attributed to the particle. The photon doesn't interact at all, whereas W and Z are caught within it and become ultra-massive. Mass is not an intrinsic property of matter, but the consequence of a dynamic interaction.

Contrary to what one might have expected, the publication of the articles did not, in its immediate aftermath, have much of an impact. To use Higgs' words: 'Our articles, initially, were completely ignored.' Then, slowly, things changed. Partly because the proposed solution appeared simple and elegant, partly because it found an exceptional sponsor, Steven Weinberg, the father of electroweak unification, who, in his articles and seminars, started citing Higgs' mechanism more and more frequently as a key element in the breaking of electroweak symmetry.

But at the precise moment when the most authoritative scientists in the world were putting the work of the generation of '64 at the centre of the Standard Model, the problems started.

Because, whilst the new particles predicted by the theory were all, systematically, being discovered, there was no trace of the particle responsible for the new field. Nobody seemed capable of finding the ghostly particle on whose existence the entire construction rested, to the extent that some started doubting that the Higgs boson even really existed.

Geneva, 8 November 2011

Even though it's my birthday, it's a morning like any other. Alarm at seven, copious Italian coffee made with my old moka, a glance at the computer to check the Cessy logbook and to be sure that 'the baby' has had a peaceful night.

I have always referred to the CMS (Compact Muon Solenoid) detector with the affectionate nickname ('bimbo') that Tuscan parents use for their children. For some time now, I have been the CMS spokesperson, which literally means mouthpiece but in practice refers to the person who is chosen to lead the collaboration of three thousand or more scientists from all around the world.

I was part of that small band of visionaries that in the early 1990s dreamed up CMS, even before designing and building it; we've always considered it to be our baby, despite the fact that the little one weighs more than fourteen thousand tons and is as tall as a four-storey building.

CMS is a modern cathedral to technology, set up in an enormous cave a hundred metres below ground, in the vicinity of the small French village of Cessy, close to Geneva. The great accelerator, the LHC, passes through right underneath here, and it's in the heart of CMS that the intersections of the beams occur. The collisions between high-energy protons are recorded by a series of concentric shells of sensors. It's a kind of gigantic, enormous digital camera, capable of identifying the particles produced in the collisions, which in turn follow one another at the mind-boggling velocity of a billion collisions per second.

It took twenty-five years to build it and get it working. We came through so many and such serious crises that it still doesn't seem real that everything functions perfectly. The generation of scientists that preceded us had been hunting the Higgs boson for decades, without success. Now it's our turn, but the fear of failure is with us every day.

We threw ourselves into the venture with the courage and recklessness of a generation of thirty-to-forty-year-olds throwing their hearts over the bar. We proposed a gigantic machine and such complex detectors that initially everyone thought we were mad. And in fact, with the technologies available in the 1990s, it would have been impossible to make the project happen. It really needed a generation of crazy, intrepid visionaries to throw themselves headlong into the work and involve thousands of young researchers from across the world. The secret was to take a leap forwards in our knowledge and devise new sensors, invent innovative materials, develop modern technologies, which might achieve the improvement in quality that we needed. In the end we were successful, but it took decades of suffering and frantic efforts.

We passed through hell, surviving all kinds of crises: sensors that didn't work, materials we were unable to produce, exploding costs, deadlines we couldn't meet and tremendous accidents, like the one which, in the early days of the LHC, damaged scores of magnets and put the accelerator out of action for more than a whole year.

But in the end, we did it. We're here. We've collected data and published so many articles. Everything is ready to hunt the Higgs. During the course of 2011, the LHC worked amazingly and gave us many more collisions than were initially predicted. It's true that Fabiola Gianotti, who leads Atlas, the other experiment, and I tormented our friends on the machine almost every day, demanding exceptional efforts so that they could give us the statistics we needed. And they listened to us. We gathered enough data to be able to begin to say something about this damned boson.

For several months, hundreds of young people have been at work analysing the data that we have recorded on disk and the work has become frenetic in the last weeks.

At the height of summer, we were on a roller coaster. At certain moments we seemed to glimpse something, then it all disappeared. Fluctuations, the tricks of statistics: for several weeks there appears to be a signal, then the area that seemed interesting empties of data and everything returns to normal.

The work of analysing the data is organized into groups, each made up of around fifteen physicists, mostly young people. They have restless, brilliant minds, and are concentrating on a particular way the Higgs decays. There are dozens working on different channels and the groups are independent of each other, only every now and then exchanging information about what was happening. But I am the spokesperson and so able to participate in all the meetings. I do this routinely, following, in particular, the two or three groups where they are studying the most promising decay channels of the Higgs. The ones that, if the damned particle really does exist, should be the first to reveal something.

And it happens to be today, the eighth of November 2011, that the two leading analysis groups are meeting on the same morning, one hour apart. As is usual, each group doesn't know what the other is doing. In the first group something happens: 'Guido, this strange excess appeared around 125 GeV.' I take note, without showing any particular enthusiasm. Then I join the second group: 'It seems to us that something is happening around 125 GeV.' So, I then call both groups together. There's no need for a lot of talk, and nobody is mentioning the word discovery, but everyone's eyes are shining. 'Maybe you've just given me the best birthday present I could ever have imagined.

But it's still too early. Now I want you to do everything you can to make it disappear.'

This is the least well-known part of our work. Outside the circle of experts, people think that when physicists find the hint of a signal, they do everything they can to amplify it, to make it clearer. In fact, the exact opposite happens; in these situations, we multiply our efforts to kill it. Because we need to be sure that it's not a malignant statistical fluctuation of the background or that we haven't made a mistake in modelling the simulations and that there haven't been any errors in the selection of events or the reconstruction software. Only at the end of this process, when we can't find a cause that can explain the signal that appeared in the data, can we stop and make the results public. Publishing the results means asking other colleagues, those who have not participated in our work, to check what we have done, for them to look for other potential sources of error.

The work to kill the signal carried out after that meeting produced no results. The anomaly resisted all checks and controls. By the end of November, it was clear that an excess of events similar to ours was also figuring in the Atlas data. And the rest is history.

The preliminary results were presented to the world at a special seminar on 13 December 2011. But the signal was still too weak for us to make a significant announcement. In the first six months of 2012, other data were gathered and, when the timid signal reappeared in both the experiments, there were no further doubts. In a second special seminar, on 4 July 2012, Atlas and CMS announced their discovery to the world.

A strange field that occupies the whole universe

The frantic race we ran to discover the Higgs left us out of breath. During and after the discovery, we were overwhelmed by unprecedented media attention. We received prizes and recognition which gave us great pleasure, but nothing compares to the emotions we felt at the first signals from the Higgs. None of us will ever forget the realization that we were among the first people to see the traces left by a particle that had been hunted for almost fifty years, nor the excitement, mingled with anxiety, that you could read in the eyes of the youngsters at the forefront of edging forwards the frontiers of knowledge.

They were such frantic and agitated months that there was no time for reflection. Only afterwards, when we were able to view all of this in a calm frame of mind, did we begin to understand what we had achieved.

The discovery of the Higgs boson is destined to leave a mark and be remembered for decades if not centuries. It's a watershed between a before and an after. Today we can describe in detail the mechanism that breaks the electroweak symmetry and differentiates among them quarks and leptons.

The fact of having succeeded in measuring with precision the mass of the Higgs allows us to reconstruct the history of the first moments of the life of our universe. Just one hundredth of a billionth of a second had elapsed after the Big Bang before the new-born universe became an immensely dense and extremely hot object, expanding at impressive velocity. Gradually, as its dimensions increase, so its temperature decreases. It already contains all the mass and energy needed to construct the gigantic universe that we know, but none of what is churning around inside it would be recognizable. It is an immense dust cloud of elementary particles, all with zero mass, chasing around everywhere at the speed of light.

But suddenly everything changes. No sooner does the temperature drop below the critical level which allows the Higgs boson to scamper freely around all over the place, like its travelling companions are doing, than a phase transition occurs; its field crystallizes into a stable state and a myriad of free

Higgs bosons remain trapped forever in this new structure. The vacuum acquires a new property; enriched by the presence of the Higgs, it becomes an electroweak vacuum, and the particles that travel through it begin to differentiate themselves on the basis of the intensity of their interaction with this new field. So, the elementary particles, as we know them today, are born; leptons and quarks become distinct, and W and Z become massive. In that moment, the mechanism that will allow matter to acquire the consistent and enduring forms that we know so well today is produced.

Stable matter, which for millennia monopolized philosophical debate, would not exist without the Higgs boson. Without its field there wouldn't be light quarks aggregating to create the first protons. Without it, not even electrons would have the little mass that marks them out and that will allow them, in due course, to orbit around the first nuclei to form atoms.

We discovered the delicate, subtle mechanism which set our universe on the path to the construction of enduring material objects: nuclei, atoms, dusts, stars, galaxies, planets and so on right down to us. Without the Higgs boson, none of this would exist. Today we know something we were unaware of until a few decades ago and which is forcing us to rewrite the physics books.

The strangeness of the Higgs and the many mysteries it hides

With the discovery of the Higgs, the Standard Model is complete. But the newcomer is a strange particle, and it's no coincidence that it's always represented as slightly apart. It doesn't belong to the family of matter particles, leptons and quarks, because it's a boson. But neither does it have much in common with other bosons, the ones that carry the interactions. Because it's a scalar means it has zero spin. It's the first fundamental scalar particle of the Standard Model; there are no others.

It's as if it were the simplest of the particles, one that could be drawn by a child: a sort of elementary prototype, no frills, just bare bones, exclusively characterized by mass. And yet it plays a fundamental role. Since it interacts with the known particles but also, if they exist, with those not yet discovered, it behaves like a kind of antenna, capable of entering into relationships with all material forms. For this reason, it is essential that we measure all its properties. For now, everything seems normal, the Higgs doesn't present significant anomalies compared to predictions, but the results do not allow us to exclude the possibility that it is concealing something strange.

For example, it could be the first representative of a whole family of other scalar particles. The Standard Model does not predict this, but nature could have some surprises in store for us. In the LHC data we could find a kind of cousin to the Higgs boson, some other members of a much vaster family, and then suddenly we would have to review the overall picture once again.

This new particle could also act as a portal for discovering other new ones, maybe weird ones or ones with bizarre behaviour: quasi stable particles which could wander slowly around our apparatus for days or weeks on end, just to decay into a spurt of familiar particles, maybe when the data acquisition is already complete. Or invisible particles, silent ghosts which pass through our detectors without leaving any trace. Being ready for surprises, taking account of the most unusual behaviours of matter, is one of the hardest things about our work.

Or it might hold in reserve for us some new information with regard to the role it played in the material constitution of our universe – a role that we already know to be fundamental, but which might have been even more relevant than we imagine.

At the origin of our universe there is a still rather mysterious phase, known as cosmic inflation. It's the phase during which a quantum fluctuation gives rise to a tiny bubble of space-time which fills with a handful of *inflatons*. These strange particles have been hypothesized to explain the dynamics of what is conventionally known as the Big Bang, the initial paroxysmal expansion which gave birth to our universe. Inflatons are scalar particles, which alone can excessively swell the minuscule portion of space-time in which they are installed. Some hypothesize that it is precisely the Higgs boson, the first fundamental scalar observed in nature, which played a role even in the very first moments of our universe. But the available data is still insufficient. To know whether it really was the Higgs boson that unleashed that inflation, further studies are necessary.

Following on from the discovery of the Higgs boson, we are living through a kind of magic moment for physics. On the one hand, we have closed a chapter that had been open for almost fifty years. Now that we have found the final particle that was missing from the roll call, the Standard Model of fundamental interactions is complete. But also, as we are celebrating another triumph for this theory, we know full well that the list of phenomena for which the model does not offer any explanation is so long as to be frankly embarrassing.

Just for starters, it doesn't include gravity, the commonest of the interactions. There are the completely unknown mechanisms that define mass and hierarchies of neutrinos; we don't know where anti-matter ended up and we still haven't managed to unify all the fundamental forces in a consistent manner.

But there is above all one property that makes the Higgs boson a very special particle. From mechanisms intrinsic to quantum mechanics, we know that each particle is perpetually surrounded by a kind of cloud of virtual particles. These are ghost particles, which are extracted from the vacuum for a very brief moment before returning to it immediately afterwards, as soon as they are reabsorbed. It is an inevitable mechanism which, in the case of the Higgs boson, would involve uncontrolled growth in its mass. We do not know of any mechanism that could protect a scalar particle from this phenomenon, and yet its mass remains there, motionless, at 125 GeV.

There just must be something. Maybe there are other massive particles, beyond those known in the Standard Model, which hover around it and cancel out its effects. Or else there is some other active mechanism which protects it from this unstoppable drift.

One way or another, the Higgs boson still seems to be hiding many secrets.

6

Shining stars and black stars

Curiosity and a desire for knowledge are at the root of scientific research. Human curiosity is perhaps something innate, a product of evolution which granted anthropomorphic apes an instrument for investigating their surrounding environment and thus to better exploit its potential. You just need to talk to a group of children about the most complex of subjects, such as how an aeroplane works or where the light of the Sun comes from, to discover the most unexpected pathways that the youngest and most flexible minds are capable of navigating and the pleasure with which they do it.

On the other hand, myths describe a desire for knowledge as an uncontrollable compulsion, so irresistible as to trigger sometimes the most terrible of catastrophes. Adam, the first man, breaks the spell of Eden to taste the fruit of the forbidden tree. He is fully aware that by breaking the tabu, he will renounce eternal life and will be condemned to suffering and death, but the impulse to bite into the forbidden fruit is stronger than any divine command.

The concept resurfaces in the Homeric myth of the Sirens, mythological creatures with the face of a woman and the body of a bird. Their beguiling song brings sailors to their ruin, and when they address Ulysses, he is incapable of resisting the idea of approaching the perilous rocks on which they live.

> Famous Odysseus, great glory of Athens, draw near, / and bring your ship to rest and listen to our voices. / No man rows past this isle in his dark ship / without hearing the honeysweet sound from our lips. / He delights in it and goes his way a wiser man. / We know all the suffering the Argives and the Trojans endured, / by the gods' will, on the wide plains of Troy. / We know everything that comes to pass on the fertile Earth. (*The Odyssey*, XII, ll. 184–191)*

Critics have often pointed out the seductive power of the Sirens' song as an irresistible element in their fatal attraction. But the text of the *Odyssey* is explicit in emphasizing the element of knowledge: 'We know everything that comes to pass on the fertile Earth.'*

It's the passion for knowledge that pushes Ulysses to challenge destiny. The concept had already been developed by Cicero who, in his philosophical dialogue *De Finibus Bonorum et Malorum* (*On the Ends of Good and Evil*), wrote: 'Apparently it was not the sweetness of their voices or the novelty and diversity of their songs, but their professions of knowledge that used to attract the passing voyagers; it was the passion for learning that kept men rooted to the Sirens' rocks.'†

The desire for knowledge as an irrepressible impulse can be found in the etymology of the word 'desire', which has given rise to different interpretations, at times at odds with each other. From the Latin *de-sidera*, to perceive the distance from the stars, it would mean not having reference points, sensing

* Quotation from Homer, *The Odyssey*, translated by A.S. Kline, CreateSpace, 2016.

† Quotation from Cicero, *On the Ends of Good and Evil*, translated by H. Rackham, Harvard University Press, 1931.

a desperate need to bridge the gap which makes us feel lost, fragile. In some accepted uses, it is the feeling caused by the suffering that comes from the separation of the earth from the stars. It is the spasmodic, desperate search for something that can overcome this feeling of being violently separated. To find in ourselves that tiny fragment of the stars that will reconnect us with the cosmos, build a bridge between the vile, corruptible earthly world in which our existence takes place, and the eternal, unchanging world of the celestial spheres.

It's an old story, repeated year on year, when millions of humans stand looking up, in the clear summer nights, around 10 August. In the darkness they are looking out for *shooting stars* and when they see the little trail left by a meteor from the Perseids, as they celebrate together, they repeat the ancient formula: 'Make a wish! It will come true!'

Very soon, in fact, we shall see that the material bonds that connect us to the stars are much tighter than, until some time ago, we might have imagined, and that every single cell of our bodies contains elements that were generated at the heart of giant stars, which disappeared since time immemorial. But what are stars really made of? What material relationship is there between us, the Earth we inhabit and what is happening in those luminous stars?

What stars are

The matter that stars are made of presents itself in a different form from normal matter. What surrounds us is made up of electrically neutral atoms, with positive protons of the nucleus equal in number to the negative electrons that rotate around them. Stars, on the other hand, are enormous concentrations of ionized gas, a new state of matter known as *plasma*. The term derives from the Greek and has the same origin as the Italian verb *plasmare* (to shape/mould), indicating something which doesn't have a predetermined shape, but that can be moulded.

Plasma is a gas in which the atoms have lost one or more electrons and have become electrically charged. The cloud of electrons torn out of the atoms moves around with them, as a result of which the volume occupied by the plasma remains electrically neutral. The matter that stars are made of is in a state that is different from the solid, liquid and gaseous states that we spoke of previously.

The properties of this fourth state of matter eluded us for a long time because in our everyday experience plasma presents itself to us in fairly transient forms, the most common being lightning bolts. The difference in electrical potential that unleashes the lightning in a thunderstorm tears the electrons out of the column of air molecules that are present between the cloud and the earth. The lightning is precisely that channel of plasma, superheated and ionized gas, which is an excellent electricity conductor.

Plasma is also what we call the ionized gas which lights up the neon tubes of commercial advertising hoardings, as well as the small cells in certain large flat-screen televisions, known for that reason as plasma TVs, and it is widely used in the superconductor industry to treat surfaces. Its properties are also being studied to create nuclear fusion power stations and to find new ways of accelerating beams of particles.

The universe is full of plasma. This material form, composed of charged particles which move freely and independently, is highly reactive, can conduct electricity and generate strong magnetic fields.

Stars are enormous spherical concentrations of plasma, inside which the gravitational pressure has set off nuclear reactions which generate energy. Every star is a battlefield on which two fundamental forces of nature clash. The weakest of all forces, gravitational pull, acting silently for hundreds of millions of years, has succeeded in concentrating an immense quantity of hydrogen and helium atoms. The primordial gases, whose nuclei were produced immediately after the Big Bang, piled up around the regions where a small density fluctuation occurred.

Gradually as the vast sphere becomes more massive, the gas of the innermost strata, compressed by the immense force of attraction, superheats and ionizes, that is to say it becomes plasma. But nothing escapes the grip of gravity which continues to press down creating abnormal pressures and temperatures. When temperatures exceed ten million degrees, nuclear fusion reactions are triggered.

Like every other star, the Sun is principally made up of hydrogen nuclei which fuse together to form helium nuclei. The latter have a mass which is less than the sum of the masses of the reagents, as a result of which an immense quantity of energy is released in the process. The innermost strata, superheated, are thrust outwards and for millions or billions of years, depending on conditions, a balance is achieved between the force of gravity which tends to compress the star and the strong force which triggers thermonuclear reactions which lead to its expansion.

The energy produced in the fusion is emitted in the form of neutrinos and photons. The former, as light as they are and little inclined to interact, escape immediately from the star's grip and move away, travelling throughout the whole universe. Photons, on the other hand, remain imprisoned at length in the high-density plasma which forms the core of the star, managing to escape only with a great effort and after infinite adventures. When they do succeed, a new star shines in the sky and illuminates the whole surrounding area with its beneficial rays.

The light emitted depends on the surface temperature, which is always high, but pales in comparison to the many million degrees of the inner strata. The *coldest* stars, so to speak, have surface temperatures of several thousand degrees and their light is reddish. The hottest stars, those which exceed tens of thousands of degrees, by contrast emit blue light.

The universe is populated by a myriad of stars, in much greater number than those visible to the naked eye. If we head away from the lights of the city, in conditions of good visibility, for example in the high mountains on a cloudless night, we can manage to see several thousand. Many do not live in isolation, like our Sun. Around half of them have one companion, forming thereby part of a binary system, and some have more than one. Almost all of them are likely to be surrounded by more or less complex planetary systems.

The dimensions of stars can vary hugely. The smallest, in order to be able to trigger thermonuclear reactions, should have a mass of at least one tenth of the Sun. The largest ones have monstrous masses, a hundred times more massive than our mother-star. Since the colossi of this category have very low density, their dimensions are inevitably impressive. Rigel, a blue super-giant in the Orion constellation and visible with the naked eye, has a radius estimated to be equal to half the distance between the Earth and the Sun. But this is nothing but an insignificant little star when placed next to UY Scuti, a red super-giant close to the centre of our galaxy which has a crazily large radius. If it were placed at the centre of our solar system, it would almost touch Saturn's orbit.

The turbulent side of the Sun

The star that illuminates our days is a small-to-medium sized star. And even though, by comparison with the Earth, it is an enormous celestial body, it is classified as a *yellow dwarf* because its surface temperature is around six thousand degrees, and it emits light mainly in the cyan-yellow range while covering all visible frequencies and part of infra-red.

The plasma of which it is composed is formed of around three quarters of hydrogen, a quarter of helium and a small percentage of heavy elements. It was born around 4.5 billion years ago from a gigantic cloud of gases and dust which was rotating around itself, having formed between the great Perseus and Sagittarius arms of our galaxy. Cooling down, it ended up collapsing. When gravity overwhelmed the impulse to expand, a wide disc of gases and dust formed rotating around the centre, where a considerable quantity of hydrogen gas was concentrated. A solar nebula was born, and all around, in the accretion disc, other smaller centres of aggregation could already be glimpsed, which would give rise to the great gaseous planets of our solar system. The moment when the gravitational pressure inside the central body triggered nuclear reactions, a new star was born.

The slow rotation of the great disc of gases and dust from which the Sun was born lies at the origin of its rotational motion, which is still evident today. Not being a rigid body, the enormous ball of white-hot plasma does not have a constant rotation period. To complete a full rotation, it takes less time at the equator than at the poles, where the motion is slower. The compact nucleus, which is at the centre, where the majority of the nuclear reactions take place, is much faster. Its rotation period is around one week.

Rotating charged particles produce magnetic fields and we shouldn't be surprised that the Sun is enveloped in these. Here the complexity of the role played by plasma comes into play. Since the rotation velocities are different at the equator and the poles, the magnetic field drawn out by the solar plasma will show significant perturbations. Moreover, inside the Sun, the plasma is subjected to massive radial convective motions: enormous portions of white-hot plasma rise up towards the surface under the impulse of the immense heat emanating from the nucleus. These movements produce other magnetic fields, which are superimposed on those generated by the rotational motions. In their turn, the movements of plasma are modified by the forces that are created when charged particles move around a magnetic field.

The result of this is a quasi-chaotic system, which develops turbulent phenomena within largely stable periodic cycles. One of these phenomena is known as sunspots. These are regions of the solar surface that are distinct from the surrounding areas by having a lower temperature and a strong intensity of local magnetic field. The magnetic activity is such as to slow down the supply of white-hot plasma from the underlying strata, as a result of which the temperature of the zone drops by more than a thousand degrees. The area concerned continues to emit light, but to a much lesser extent by comparison with surrounding areas, and thus it appears as a dark spot. Sunspots can cover areas equivalent to twice the surface of the Earth and generally last a few weeks, before being reabsorbed and returning to normality.

The number of spots that appear on the Sun is an indicator of solar activity and follows a cyclical pattern, reaching a maximum, on average, every eleven years. The dynamic of this mechanism has not yet been fully understood, just as, to date, there are no detailed models to describe solar flares and the formation of prominences.

Surface eruptions, or solar flares, are impressive releases of energy which are thought to arise from the reconnection of lines of force of the strongly perturbed magnetic fields which develop on the surface of the Sun. They often appear in the proximity of sunspots and most frequently at the peak of a cycle of solar activity. The emission of great quantities of plasma into space can bring with it very strong solar winds: clouds of ionizing particles, expelled at great velocity, which can reach Earth, interfere with its magnetic field and penetrate its atmosphere. These phenomena are particularly dangerous for electrical grids and telecommunications because they are capable of irreparably compromising the satellites which orbit our planet and, in the most serious cases, can cause biological damage to living organisms.

One of the most spectacular events, the first to attract the attention of scientists, was recorded over a century ago. It's remembered as the Carrington Event, after the British astronomer Richard Carrington who observed it, almost by chance, on 1 September 1859. The violent solar eruption that generated it produced a cloud of charged particles which quickly reached Earth, with consequent aurorae borealis visible as far south as Rome and Cuba. The geomagnetic storm produced by the interaction with Earth's magnetic field sent telegraph lines haywire. If a similarly violent event occurred today, it would cause electrical blackouts across many continents and would damage, perhaps irreversibly, around half the satellites we use for telecommunications.

Solar flares are often accompanied by solar rings and prominences, emissions of great quantities of plasma in the form of arcades or filaments, which can spread over hundreds of thousands of kilometres. And so, our Sun is still hiding a number of mysteries. Many details around its activity and the origin of its periodic cycles are still unknown to us. We don't know about the mechanism that produces its corona, the enigmatic highly energetic sheath of gases and particles which extends hundreds of thousands of kilometres into space and whose temperature can exceed a million degrees. We do not know the origin of solar flares and prominences, even if the principal suspect is still the extremely complicated magnetic field of the Sun and its multiform reactions with the white-hot material of which it is composed. The only star that we can study up close continues to reveal surprising features. Beneath the peaceful and reassuring appearance of the placid Sun lurk terrifying turbulences.

The nocturnal vision of the stars has inspired poets of every civilization and has calmed our anxiety. But when we observe these wonderful stars in detail, we discover environments dominated by catastrophes of unimaginable dimensions. And yet this is nothing compared to the phenomena that are unleashed when stars reach the end of their existence.

The spectacular end of a tranquil star

The destiny of a star is written in its size and the particulars of its composition. The length of its life depends on the type and speed of the thermonuclear reactions that take place inside it.

The first stars to light up in the primordial universe were very different from the ones that shine in our firmament today. They have been named *megastars*, precisely because they were giant stars, hundreds of times more massive than our Sun, and made up solely of hydrogen and helium, the only available elements. The nuclear fuel of such large stars is used up more quickly, because the enormous quantity of material exerts such pressure on the inner strata as to heat the nucleus to terrifying temperatures. In the heaviest stars a billion degrees is exceeded by some measure. In these conditions, nuclear fusion reactions proceed at great speed and the available fuel runs out in the matter of a few million years. The first stars lived ephemeral existences compared to the billions of years of the stars of successive generations, like our Sun.

When the temperatures of the nucleus are so high, the fusion reactions involve all elements. As soon as hydrogen and helium run out, heavier elements – carbon, nitrogen and oxygen – start to fuse. When the point is reached where silicon nuclei are fused and iron is generated, the process stops. Further reactions are not possible and the core of the star collapses catastrophically. Clouds of residual gas, rich in heavy elements, are expelled at great velocity and cover remarkable distances. It is from these that other stars are created, which will eventually feed the whole process.

Our Sun was preceded by many generations of stars which, through their death, enriched the original nebula with hydrogen and helium, but also with heavy elements which aggregated to form rocky planets. There would be no Earth, no water, no air, no living forms based on the carbon cycle, without this interminable sequence of the life and death of stars. Every atom of our bodies is the result of such a long process. The nuclei of the iron in our haemoglobin, of the calcium in our bones, of the oxygen and carbon of the organic molecules which compose the fabric of which we are made, all derive from the immense nuclear furnaces which preceded the Sun.

As he chased Adam out of the Earthly Paradise, the God of the Bible pronounced the most famous of all curses: 'For dust thou art and unto dust shalt thou return.' The terrible condemnation to mortality and suffering might appear less harsh if we take cognizance of the fact that, yes, we are dust but, to be precise, star dust. Something profound unites us with the most luminous stars which dot the night sky.

Scientists consider that our Sun is living the best years of its life, roughly in the middle of its existence, and that it might continue to shine for another four or five billion years. Dwarf stars, like the Sun, are fortunate, because their nuclear fuel burns so slowly that they can live a very long time. However, even for them, sooner or later, the final moment will come.

In the final phases of the cycle of its life, after having burnt up all the hydrogen in the nucleus, the Sun will swell up excessively and turn into a red giant. When the nuclear reactions cease, gravitational force will crush it and make it contract. Compressed into a smaller volume, the nucleus will superheat and set off nuclear reactions in the shell of plasma which surrounds it. The increase in temperature will produce reactions that will follow each other at a frenzied velocity. The star will become hundreds of times brighter, even though the light emitted will be redder. To attempt to disperse the immense heat, which is agitating it, the dying Sun will dilate, expanding its outer strata to excess. In five billion years, or thereabouts, it will swell up hundreds of times, eventually vaporizing all the inner planets, including Earth.

This event shouldn't arouse too much concern, because when this happens our planet will not have been home to any form of life for some time. Disappearing into a cloud of gas will be a burning wasteland. The water of the oceans and the terrestrial atmosphere will already have been ripped away by the solar wind and by the terrifying increase in the Sun's brightness.

The following phase will be no less spectacular. The great red star will continue to shine for a very long time, maybe a billion years, then, when the fuel from the shell has run out, the nucleus will contract, eventually stabilizing with a diameter similar to that of the Earth. At that point the Sun will have become a *white dwarf*, a miniature star, a barely bright spot at the centre of an immense shell of gases and dust.

We have just been describing the end of stars that have masses smaller than or similar to the Sun. As the vast majority of known stars belong to this category, becoming a white dwarf will be the fate of 97 per cent of the stars in our galaxy.

The name recalls the tiny dimensions of the little celestial body, by now reaching the final phases of its existence. The light emanating from it is white, because it covers all the visible frequencies, but it no longer stems from nuclear reactions, which by now have ceased forever, but from other mechanisms.

Despite having the dimensions of a planet, its mass is enormous; it still contains around half the matter that made up the Sun, which means that its density is very high. If we were able to extract from a white dwarf a small portion of matter, equal, for example, to a teaspoonful, we'd need a large HGV to transport it, because it would weigh several tons.

The matter of which it is composed is made up of inert nuclei, carbon and oxygen, the final products of the Sun's nuclear reactions, mingled with an electron gas. The gravitational pressure is so ferocious that everything ends up so compacted as to bring quantum effects into play.

The density of matter in the core of the mini star increases so much that too many electrons crowd into the same points. Following Pauli's exclusion principle, they can't find themselves in the same place as other fermions, because they would have the same quantum number. If they are located in a determined portion of space, which is already populated by other electrons, they must increase their velocity in order to enter a more energetic quantum state. The greater the number of electrons going to gather in the same position, the higher the velocity the newcomers need to achieve.

In this way the electron gas gains energy, as the density increases, and ends up exerting pressure which contrasts with every eventual contraction. When the impulse for expansion, due to the pressure of the electrons, equals the compression due to gravity, the dimensions of the small celestial body stabilize. All of which demonstrates, once again, that quantum mechanics not only provides the key to understanding the behaviour of matter on a microscopic scale but it is also indispensable for an understanding of the dynamics of the macroscopic world. Without it, it would be impossible to understand what is happening to billions of stars reaching the end of their existence. The pressure, which denies white dwarfs any eventual reduction in volume, and which derives from Pauli's exclusion principle, is called *degeneracy pressure* and the strange matter they are composed of is referred to by the name *degenerate matter*.

The equilibrium that governs the dimensions of the white dwarf, which will be born from the death of the Sun, could last hundreds of billions of years. As soon as it is formed, the mini star will have a surface temperature higher than a hundred million degrees and will emit photons into the surrounding space by radiation. Slowly, very slowly, its temperature will drop, and the moment will come when it will no longer be a luminous object. Its destiny is to become a *brown dwarf*, a tiny, dark object, too cold to radiate, which will travel, unobserved, through the darkest shadows. But this might take several hundreds of billions of years, a considerably longer period of time than the current life of the universe.

Supernovas and neutron stars

The fate of the most massive stars is, if possible, even more spectacular; almost as if, with a formidable finale, they are compensating for a much shorter existence than that of their smaller sisters.

The life of a giant star, weighing ten times heavier than the Sun, is much shorter because the nuclear reactions that take place inside the nucleus are extremely rapid and everything fuses more quickly. Just as with the megastars, the extremely high temperature of the innermost strata triggers a complicated chain of nuclear reactions, which involves increasingly heavier elements. Then, suddenly, when things get as far as iron, everything stops. At that point the central nucleus implodes. Crushed by the immense gravitational force, which is no longer offset by the heat, it collapses catastrophically. It reduces by several thousand times in size and all the upper strata find themselves suspended in the void. Deprived of support and drawn towards the centre of the still imposing mass of the nucleus, they hurtle towards the small compact body. The impact is devastating and unleashes further nuclear reactions which smash the star to pieces in a blinding flash of light. We have just witnessed the explosion of a *supernova*, one of the most catastrophic events in the universe.

The term supernova is an intensifier of *nova*, the Latin adjective which was used to describe the birth of a new star. The astronomers of the past, seeing a new light, would imagine that a new star had been born. The term has survived, even though today we know that those blazes of light are quite the opposite of what used to be thought and signal the final moments of a great star.

Over a certain length of time, which can be anything from a few weeks to several months, a supernova emits the energy that the Sun will radiate in the course of its entire existence. The explosion expels the majority of the material which made up the parent star. An enormous mass of gases and dust, equivalent to many solar masses, spreads through the interstellar medium at speeds that can reach a tenth of the speed of light. The immense quantities of material moving at such speed produce a massive shock wave, which in turn triggers other phenomena. But the most mind-blowing thing is what happens in what remains of the nucleus.

A large mass, around one and a half times the Sun's mass, is still concentrated in this small region. The gravitational force is so powerful that the electron pressure that held the white dwarfs in balance no longer works. To compensate for the impulse towards contraction, the electrons that are agitated so as not to violate Pauli's exclusion principle, should, in this case, move at velocities greater than the speed of light. Since this isn't possible, a change of state occurs. The nuclei fall apart reducing to a 'pulp' of protons and neutrons among which electrons are moving extremely fast. But gravity, which by this point nothing can counteract anymore, fuses electrons with protons and transforms them into neutrons. This reaction releases a vast flow of neutrinos which hasten at speed to inform the entire universe of the end of a great star. In a few dozen seconds the diameter of the nucleus has contracted by several thousands of kilometres to something like ten kilometres. And a *neutron star* has been born.

These are very small astronomical objects comparable to an immense atomic nucleus, made up above all of neutrons and held together by gravity. Their density is frightening, a hundred million times greater than that of the degenerate matter of white dwarfs, and their heat is extremely intense, around ten million degrees. In this case not even the most solid parts of the Earth's crust would be capable of supporting the weight of a teaspoonful of the matter which makes up a neutron star.

The process that holds them together resembles that of the white dwarfs. In this case it's the neutrons exerting pressure that counteracts the grip of gravitation. Having fractional spin, they too are fermions, and when they are confined in too restricted a space, they increase their velocity and produce the pressure that keeps the small, ultra-compact body in equilibrium. Neutron stars too are made up of degenerate matter, which is much denser and more compact than that of white dwarfs.

We do not really know what happens in the core of these bizarre celestial bodies. Beneath a very hard crust, which we guess to be a couple of kilometres thick, one should find that kind of *neutron 'pie'*, which produces the degeneracy pressure. There is speculation that in the innermost and densest strata of the nucleus there are concentrations of even more exotic matter: strange sub-atomic particles, plasma of quark and gluons, states similar to those recorded in the particle accelerators, or even shells of quarks. But for now, these are merely speculations.

Neutron stars rotate furiously around themselves. The origin of this movement arises from the motion around the parent star's own axis. By the law of conservation of angular momentum, when the great star implodes, the rotational velocity must per force increase – a little like happens to ice-skaters who increase their rotational velocity by drawing in their arms. Since the reduction in dimensions is impressive, so is the increase in velocity. If we were to compress the Sun, with its rotational period of around a month, into a sphere ten kilometres in diameter, it would complete more than a thousand rotations per second. And, in fact, neutron stars have been discovered rotating around themselves dozens or hundreds of times per second. In time, because of the energy emitted during the rotation, the velocity reduces.

With the implosion of the parent star, the initial magnetic field, now concentrated into a vastly smaller volume, is also amplified disproportionately. Neutron stars are accompanied by magnetic fields billions of times greater than that of the Earth. Some are so excessive that they would be lethal for humans even thousands of kilometres away.

When the magnetic axis doesn't coincide with its rotational axis, the neutron star emits from its poles a powerful beam of electromagnetic waves. Viewed by an observer on Earth, it resembles a kind of radio beacon that sends out signals at regular intervals, each time the radiation cone hits our planet. This is the behaviour of *pulsars*, a pulsating radio source, as they were called before it was understood that they were neutron stars. Their radio emissions are so regular that, initially, astronomers had even thought that they might have been produced by alien civilizations.

But these very special stars have also given us an even more important gift. Thanks to them, a mystery, which scientists have been racking their brains over for decades, has been resolved. We saw earlier that it was explosions of supernovas which enriched areas of the universe with metallic elements. But nuclear reactions stop at iron. We still had to understand where the heaviest elements of the periodic table – gold, platinum, lead and so on up to uranium – had come from. We know that they cannot form inside the great furnaces operating in the core of the great stars. And so, who produced the gold that has adorned kings, warriors and princesses since earliest times?

The answer has been supplied by multi-messenger astronomy. This is the name for the new discipline in which the same astronomical phenomenon is studied using all the signals emitted during the process. In one specific case, at 12.41 on 17 August 2017, the massive Ligo and Virgo devices had detected a signal of gravitational waves due to the fusion of two neutron stars. The same phenomenon, two seconds later, had produced a powerful gamma-ray burst, very high energy photons, detected by Fermi, a satellite dedicated to the Italian physicist, in orbit around the Earth. Pointing all their telescopes in that direction, astronomers were able to study the phenomenon by using the entire range of frequencies: from visible light to X-rays, from infra-red to radio waves.

The zone where the collision occurred was kept under observation for months, and when astronomers analysed the cloud of detritus produced by the event, they recognized unmistakable signs of the presence of heavy elements. Estimates speak of a cloud containing quantities of gold and platinum equal to ten times the mass of the Earth. Essentially, people knew that quantities of heavy metals could be produced during the explosion of a supernova, but now we had proof of a much more significant production. It is thanks to the collision of neutron stars that the nebula out of which our solar system originated were enriched with these metals, traces of which we now find in the surface sections of the Earth's crust and which we are able to exploit through mining.

So, it is definitively true that we are *children of the stars*, but it's specifically thanks to neutron stars and their collisions that, since time immemorial, we have been able to adorn ourselves with those precious metals that still enchant us today.

Black holes

When stars of truly disproportionate mass die, an even more surprising phenomenon occurs. If, after its explosion into a supernova, the residual nucleus exceeds three solar masses, not even the core 'neutron pie' can preserve the balance. The grip of gravity is so implacable that it crumbles the neutrons themselves, reducing to a pulp even the quarks and gluons that make them up: from the death of the star a *black hole* is born.

Black holes are truly extraordinary objects. For decades it was thought that they didn't really exist, that they were simple mathematical curiosities. And, in fact, nobody could imagine a physical mechanism capable of concentrating a mass so great as to prevent light from escaping in such a restricted space. The escape velocity from the Earth, that is the speed of rockets putting satellites into orbit, is 11 km/s. To free oneself definitively from the attraction of the Sun and escape its gravitational field, one would have to move at more than 600 km/s. But if you get too close to a black hole you are trapped forever, because the escape velocity is greater than the speed of light and nothing, not even a photon, can escape the grip of its gravity.

In fact, black holes are not 'holes', or zones void of matter, at all. On the contrary, they are points at which the density of matter explodes. They are called 'black', because not even light can escape their gravitational field. When photons, the lightest and fastest of the particles, try to move away, they end up falling back into the most profound darkness, like stones thrown up in the air.

Nobody is yet able to explain the form assumed by matter swallowed up by a black hole. The space available is so small that it wouldn't be sufficient to contain its mass even if everything was reduced to a dense core of quarks tightly packed in against each other. The question of the organization of matter in black holes touches on concepts that are so incredible as to border on the absurd.

For example, let's imagine penetrating deep inside the Sun without suffering any damage; we would experience a gradual weakening of the gravitational pull the closer we got to the centre. Leaving behind the part of the mass concentrated in the outer shells, the gravitational force would reduce; when we arrive at the centre, it would be zero. In black holes, the opposite occurs. As you get closer and closer to the core of this strange object, the force of attraction increases; it becomes monstrously huge when you find yourself at an infinitesimal distance from its centre.

Black holes are not great balls of matter like the other stars; on the contrary, they are empty spheres surrounding a *gravitational singularity*, that is a region of essentially infinite density. They are space-time regions in which all matter is concentrated in one point. But the gravitational field that they produce is so intense that that void, that kind of star made of pure geometry, is so imbued with energy that it has devastating effects on everything that approaches it.

In recent years, it has been discovered that black holes are real astronomical objects, which populate our universe. Many dozens have been identified, particularly since we have been able to record the gravitational waves emitted by their fusion. Today it is thought that, although much rarer than ordinary stars, their population is numerically significant. They constitute a whole new class of celestial bodies, with very particular properties.

Very recently it was noted that, in addition to stellar black holes, which it is thought arise from the death of large stars, there are real monsters with masses which are so disproportionate as to exclude *a priori* that they could have been formed by a star. They are the *supermassive black holes*, like Sagittarius-A*, which is located at the centre of our Milky Way, and which weighs as much as four million Suns. There are even some which concentrate in one point the mass of an entire small galaxy: billions of stars. Their dimensions are so enormous that we have no idea how they were formed.

Supermassive black holes, of mass between one million and ten billion solar masses, are found at the centre of almost all galaxies and, in some cases, provide a spectacle. One can see nuclei of galaxies which emit extremely powerful streams of radiation on every wavelength, flashes of light, radio waves, X-rays and gamma rays. Sometimes, enormous filaments of matter erupt and spread throughout the surrounding space for tens or hundreds of light years. It is supermassive black holes that produce some of the most energetic and surprising phenomena in the cosmos.

Black holes are not *bottomless pits* from which nothing emerges, as they are habitually described. When a star or a cloud of gas gets too close, their matter starts to be torn apart and to hurtle towards the black hole. As it approaches, it spirals and forms a large accretion disc which immediately becomes luminous because it is irradiating on various wavelengths. The crumbled fragments collide with each other, the friction superheats them, and everything ionizes. The plasma rotating dizzily around the centre of attraction produces strong magnetic fields, which in turn interfere with the movement of the new shreds of matter hurtling towards the black hole.

The mechanism that fuels them is not yet known precisely, but enormous jets of charged particles often appear, emitted by the black hole in the direction of the polar axes. In the jets produced by supermassive black holes, velocities have been measured that are comparable to those of light and great filaments of matter have been observed moving away from the parent galaxy for millions of light years. In this way, it was discovered that supermassive black holes contribute to the distribution of ionized matter into the great intergalactic space; and this, in turn, may be intercepted by other galaxies, thus nurturing their growth.

And so, once we understood that, by swallowing large quantities of material, black holes can become the most luminous objects in the universe, we now discover that what seemed to be deathly stars, bottomless pits swallowing up everything, in fact scatter fine matter throughout the entire intergalactic space and make it fertile.

7

The obscure forms of matter that populate the universe

The matter that makes up ordinary stars and dwarf stars, neutron stars and black holes is merely a tiny fragment of the total mass of the universe. Scientists' estimates vary, depending on the different hypotheses concerning the size of the population of black holes hiding in the galaxies. But they are all in agreement that the percentage in any case is insignificant, somewhere between 0.5 and 1.1 per cent of the total.

This seems incredible to us, because it was precisely those shining stars that suggested to us the presence of a much vaster world than the one in which our existence takes place. Moreover, if we look at the closest star to us, we discover that the mass of the Sun contains over 99 per cent of the mass of the entire solar system. And so, we're used to considering the stars to be the most massive and gigantic objects around us, and it's difficult to imagine that the universe might contain non-luminous matter which is so abundant as to render their contribution insignificant.

And yet it's not celestial bodies, whether the luminous ones which dot the skies of all galaxies or the dark, invisible ones which move around in the shadiest zones, that dominate the material structure of our universe. On the contrary, the vast majority of the matter of which it is composed presents in different forms, some of them surprising and surrounded by a halo of mystery. But let's proceed step by step.

Mostly gas and a pinch of dust

The distances that separate the stars in a single galaxy are enormous. Proxima Centauri, the star closest to us, is four light years away. If we wanted to reach the centre of the Milky Way, the journey time would be measured in many thousands of light years. And that would still be nothing compared with the distance we'd have to cover to reach the closest galaxy, the gigantic Andromeda galaxy, which is approaching us at high velocity, but is still more than two million light years away.

Galaxies are imposing structures. Their volume is enormous, but only a negligible portion of that is occupied by stars and other celestial bodies. Between one star and another is the void, which, however, is not absolute, but there is present within it a subtle and impalpable form of matter. The density of what is known as the *interstellar medium* is ridiculously low, but the volume in which it is distributed is gigantic, because of which, in the end, its contribution proves to be dominant.

The space between the stars is occupied by a very tenuous medium, containing essentially molecular hydrogen and helium which float freely throughout the whole volume. For the most part this is invisible because its density is too low. If it were possible to rake up the atoms which make up the interstellar medium, only a few would be collected per cubic centimetre in the regions where the gas is most rarefied, and a few hundred per cubic centimetre in those regions where it is densest. The only zones where the presence of material can be detected are those where large quantities of ultrafine dust are concentrated. This is the case, for example, for the central nucleus of our galaxy, which is surrounded by a kind of immense cloud; or the zones that contain the gigantic molecular clouds among which new stars can be glimpsed in the process of formation; or those that encircle a supernova which exploded centuries ago, which appears to us surrounded by its marvellous halo of residue. In each case, the dust is a wholly secondary component of the interstellar medium, given that it contributes merely 1 per cent to the matter it contains. The rest is almost all gas: three quarters hydrogen, one quarter helium. It is a residue of the great cloud of light elements produced in the moments following the Big Bang, from whose concentration the first stars were born. This considerable reserve is even now the base material from which new stars originate. In the gas, traces of heavier elements like calcium and even complex organic and inorganic molecules can also be found.

If we now look at the great intergalactic spaces, the immense volumes of void which separate one galaxy from another, we won't be astonished to discover that extremely rarefied matter can be found there too. Despite its density being millions of times lower than that of the gas which occupies the interstellar medium, its volume is so excessive that the contribution of the intergalactic medium to the total mass of the universe is absolutely dominant. It contains ionized gas, mostly hydrogen, an extremely low density but often white-hot plasma; since it emits radiation, it can be viewed with the appropriate instruments. And this is how it was discovered that the apparently empty regions which separate the galaxies, are in fact criss-crossed by a dense network of filaments of rarefied plasma. A fine web seems to wrap the galaxies and connect them to the great empty spaces that surround them.

When we take stock, we discover that 4 per cent of the total mass of the universe is made up of this fine fabric of hydrogen and helium which surrounds intergalactic space and makes its way into the galaxies themselves.

Since the remainder, a contribution of between 0.5 and 1 per cent, comes from the stars, which in turn are made up of hydrogen and helium, it is clear that these two elements predominate in the chemical composition of our universe. More than 98 per cent of the ordinary matter that composes everything that surrounds us is made up of hydrogen and helium. In that miserable 2 per cent residue we find all the rest, all the heavier elements starting with carbon, oxygen, silicon, iron and so on. And so, everything that assails us and affects us closely, as inhabitants of a small rocky planet, belongs to the category of ultra-rare elements.

But if all known matter, the matter made up of atoms, the plasma of which the stars are composed and the rarefied gases which dominate the great empty spaces, contributes merely around 5 per cent of the total mass of the universe, what is the rest made up of? What form does 95 per cent of the matter which makes up this immense structure which hosts us take?

Milkomeda and the Black Eye Galaxy

The great agglomerations of stars that we call galaxies populate the entire universe. Each one of these enormous cosmic clumps groups together tens or hundreds of billions of stars and other celestial bodies. They are distributed throughout, and their forms are diverse, but they can be grouped into three large families.

Elliptical galaxies have the shape of a more or less flattened ellipsoid. Some present as similar to enormous spheroidal globes. They are less luminous than spiral galaxies and so it's difficult to detect them, as a result of which there is a tendency to underestimate their number. Astronomers, however, think that they make up around one third of all galaxies.

The diversity of elliptical galaxies as far as mass and dimensions are concerned is extremely vast, ranging from dwarf ones, containing a few million stars, to the giants in the category, which are home to thousands of billions. The distinctive properties of elliptical galaxies are that they lack almost any dust and gas and have a population of old and cold stars. It is thought that some time ago they used up the fuel necessary to produce new generations of stars and that they might continue to shine in the sky for just a few billion years, so to speak.

By using sophisticated computer simulations people have become convinced that they are the end result of multiple collisions between spiral galaxies. Maybe becoming part of an elliptical galaxy is the fate which awaits us in the long term. Our Milky Way is destined to collide with the great Andromeda Galaxy in four or five billion years, more or less when the Sun will exhaust its supply of fuel and will expand to become a red giant. The fusion of the two galaxies will take a long time to achieve and maybe, in the end, a large elliptical galaxy will be formed, a super-galaxy which astronomers have already named Milkomeda.

The event will certainly be spectacular, even if this genuine cosmic embrace lasts billions of years. There may even be a highly pyrotechnical finale, of the kind used in some countries when an important event needs celebrating. If the collision brings the two supermassive black holes that are snoozing at the centre of the two galaxies close enough together, they too will end up fusing into a single body of disproportionate mass. The gravitational waves that will be unleashed will be billions of times more powerful than those recorded by Ligo and Virgo on planet Earth, and if the new monster starts to swallow up dust and stars in its vicinity, giving birth to an active galactic nucleus, well, then we shall see some truly amazing things.

The Milky Way and Andromeda are two fabulous spiral galaxies. They belong to the most numerous family, in which around 60 per cent of all galaxies are grouped. They are flat disc-shaped agglomerations of stars in which spiral arms can be seen slowly rotating around a more or less pronounced central nucleus. Around two thirds of spiral galaxies have a bar-shaped structure extending from the central nucleus, like our Milky Way.

Spiral galaxies have masses of between one and a hundred billion stars and their average diameter is around 70,000 light years. They generally contain either stars that have already reached the final phase of their existence, or very young stars or even those still forming. Their interior is abundant with gas and dust.

The third family, irregular galaxies, contributes around 10 per cent to the total number of galaxies, but contains those with the weirdest forms. There are some shaped like a sombrero, others which resemble a penguin, others still for which astronomers have indulged themselves in finding names that draw attention to their appearance. This is true, for example, of the Black Eye Galaxy.

Their irregular shape often arises from the interaction with other nearby galaxies, or from catastrophic internal phenomena that have distorted their structure. For some there is speculation about a colossal explosion that took place in the nucleus, and which dispersed the stars and the rest of the material to great distances from the centre. Often the distortion arises from the gravity exerted by some other nearby galaxy; sometimes, in fact, irregular galaxies find themselves in the vicinity of a great spiral galaxy of which they form a kind of satellite. This is the case with the Large and Small Magellanic Clouds, the two irregular galaxies which are found around 180,000 light years from the Milky Way and which are among the very few galaxies visible with the naked eye from Earth, obviously when conditions are particularly favourable.

These irregular galaxies are very rich in gases and dust and contain great populations of new stars. Inside them, young and immensely massive stars can be found, emitting light in the highest frequencies in the spectrum, typically blue.

As we have seen, there are a number of differences between the various families of galaxies, but there is one deep mystery that they have in common. They are all hiding something. The luminous matter, the matter that their stars are composed of, together with the large volumes of rarefied gas and dust that they contain, represents only a minimal part of the matter of which they are made up. Holding them together is something absolutely unknown: a form of matter that does not emit light and whose origin is, for now, a true mystery.

The dark side of matter

One of the most important discoveries of the last century was made by the American astronomer, Vera Rubin, who, in the mid-1970s, managed to measure the velocity of the peripheral stars in the spiral galaxies with great precision.

The laws of universal gravitation tell us that close to the galactic centre the velocity should increase with distance, then reach a maximum value and finally diminish drastically for the most distant stars. Once we know the distribution of matter, stars, dust and interstellar gas in the galaxy, we can calculate the force of attraction at a given distance and extract from this the velocity of the stars orbiting in that region.

But Rubin's observations offer a surprising result, which contrasts radically with theoretical predictions. At a great distance the velocity remains constant. The peripheral stars of a spiral galaxy, which are found maybe somewhere around its outer edge, rotate at the same velocity as those closest to the centre. It's an absurd, inexplicable result, unless we hypothesize that an enormous quantity of invisible matter is hiding in the galaxy. According to Rubin, galaxies should contain between five and ten times as much matter as we had thought. Her data confirmed the hypothesis about the presence of dark matter in the universe. The thesis had been put forward in the 1930s by the brilliant Swiss astronomer, Fritz Zwicky, to explain the dynamics of the galaxies in clusters.

The result was so astonishing that, initially, it provoked controversial reactions. I can imagine the comments of the most famous scientists: 'What?! So, we already have a clear understanding of the origin of the universe; the Big Bang theory enables us to explain the formation of the primal elements, of stars and galaxies, and then this woman shows up and turns everything that we know on its head. She tells us that the universe is full of a form of matter which is so abundant as to make everything that we had imagined up till now derisory.' Rubin was a fairly unknown scientist and, before this discovery, had published some contested, not to say inaccurate, findings.

But all the verifications merely confirmed the argument put forward by this bold astronomer. And not only that; when people began analysing the data coming from the clusters of galaxies that had attracted Zwicky's attention, further validation emerged.

Galaxies do not like to live in isolation; they are grouped together in a well-structured system of families, linked to each other by gravitational attraction. For example, the Milky Way is part of a local group of galaxies which includes around seventy members distributed over a distance of ten million light years. Around sixty million light years away is the Virgo cluster, which incorporates around 1,500 galaxies. Together they form part of the Virgo super-cluster, an enormous widely scattered family which contains more than a hundred groups and galaxy clusters.

By studying these vast conglomerates, we can repeat the exercise carried out by Rubin, extending it to the movement of galaxies. And here, too, we discover that luminous matter is broadly insufficient to explain the velocity with which galaxies move within a cluster. If we don't hypothesize the presence of a dark form of matter holding them together, all galaxies should disintegrate, and clusters should have fallen apart since time immemorial.

In the last fifty years, an impressive quantity of data has been collected about the presence of dark matter in the universe. To ever more precise measurements of the velocity of stars within spiral galaxies and of galaxies in clusters, has been added evidence obtained from the study of cosmic radiation and deriving from the use of gravitational lenses. The only thing that remains incomprehensible to everyone is why the Nobel Committee has never recognized the pioneering work of Vera Rubin in this field with a more than well-deserved prize.

The light which sees dark matter

When Dante Alighieri has to find words to describe Paradise, he finds himself faced with an absurdity. He has to represent in poetry the kingdom of the immaterial, the invisible. The challenge is to tell through images what cannot be expressed in words. And so, he adopts a brilliant strategy. The souls of the blessed appear as lights, radiances and flames integrated into an ever more indeterminate and luminous environment. Notably, to describe the Empyrean, the highest of the heavens, he uses the most impalpable of substances: light. The most blazing of the celestial spheres, that non-place beyond time and space, becomes the 'heaven of pure light' (*Paradise*, Canto 30, line 39).*

In the highest realm of Paradise, Dante celebrates the triumph of light. The amphitheatre of the blessed, the circles of the angels, even the ultra-luminous point where God himself shines, everything is made of light. When I reread these

* Quotation from Dante Alighieri, *The Divine Comedy*, translated by Allen Mandelbaum, Everyman, 1995.

wonderful lines with the eyes of a modern scientist, I am still enchanted, because a large part of our knowledge about the origin of everything emerges from a detailed study of cosmic background radiation.

It is a sea of low energy photons, which pervades the entire universe. It is an ancient residue of the Big Bang, which, in its properties, still retains the memory of that magic era when photons and matter were indissolubly connected. It took 380,000 years of expansion and cooling of the primordial universe to reach the critical temperature that allowed electrons to bond to protons and form the first hydrogen atoms. At that moment, the myriad of light particles, trapped in matter until an instant before, began chasing around the whole universe as they still do today, billions of years later.

This flow of low energy photons which floods us, incessantly and from all directions, contains a frighteningly massive amount of information about our remotest history. By developing new instruments, we have succeeded in recording every detail of this, and this has allowed us to decipher its innumerable secrets.

It's not Dante's triumphant light, but a subtle glimmer, invisible to our eyes. A weak flickering of electromagnetic waves, attenuated by the 13.8 billion years that have passed since the Big Bang and stretched towards the low frequencies by the unstoppable expansion of space-time. But it is this tide of radiation flooding the whole universe that tells us about the origin of the world. It was predicted in 1948 by the physicists George Gamow, Ralph Alpher and Robert Herman, and measured for the first time in 1965 by Arno Penzias and Robert Wilson, who were awarded the 1978 Nobel Prize for their discovery.

The detailed study of the properties of this radiation tells us much about the composition of our universe. By interacting with all the matter in the cosmos, it retains slight traces of everything that passes through it. In cosmic background radiation there remain obvious signs of the network of dark matter which holds galaxies and clusters together and this allows us to quantify its presence and understand its properties.

There is finally a third way of hunting down this mysterious component and, here too, it has to do with light. This time, however, it is precisely the visible light emitted by galaxies that is used. The technique is based on the fact that even light can be deflected by great concentrations of matter. The effect is called *gravitational lensing* and is another of the phenomena predicted by Einstein's general relativity.

Everybody has had the experience of throwing a stone up in the air and watching it fall back to the ground again after tracing a parabola. When, on the other hand, we switch on a small laser beam, for example one of those that are used to highlight something during a presentation, we have the impression that the light travels straight towards the screen. In fact, light falls too; it is attracted to the Earth like the stone, but its deflection is so minimal as to seem completely negligible. For these deflections to be significant, the gravitational attraction needs to be enormous, such as, for example, the Sun's. Light spreads along space-time lines, and if these are curved significantly by the presence of a large mass in the vicinity, that's when we will see light travelling in curvilinear orbits.

It was precisely this phenomenon that made Einstein's new theory popular. His predictions were shown to be accurate, for the first time in 1919, thanks to a historic experiment led by Sir Arthur Eddington. Making the most of a total eclipse, the British astronomer managed to observe the stars closest to the Sun and proved that their apparent position was slightly altered, specifically because the gravity of our star deflects the rays of light which pass close by. It's the same phenomenon that we observed when we spoke of black holes. If the gravity is sufficiently strong, light can orbit one of these strange bodies in perpetuity, just like the artificial satellites in orbit around our planet.

The gravitational effects on the transmission of light also lie behind the artifacts, if not actual mirages, which fill the images reproduced by the most powerful telescopes. Stars or galaxies which appear where they shouldn't be, multiple images of the same object, galaxies which change shape as if at the mercy of something that is squashing or twisting them. Gravity acts on rays of light too, deflecting them just as a lens does.

Astronomers and astrophysicists have managed to transform all of this into an investigative tool. By measuring the deflection of light, it's possible to calculate the gravitational mass which produced the space-time distortion and if this doesn't correspond to the luminous mass clearly visible with telescopes, then the difference can be attributed to the concentration of dark matter.

This exercise has been repeated thousands of times and has been applied to stars, galaxies and clusters of galaxies. Each time a new, more powerful telescope has been used, ever more precise measurements of this mysterious matter have been obtained.

The agreement between the estimates produced by the different methods is impressive. Dark matter is a dominant component by comparison with luminous matter. We have previously seen that the universe contains around 5 per cent of so-called ordinary matter, known matter which assumes the form of gases and dust and stars of all kinds. Now we have to take into account the fact that dark matter is much more abundant; more than a quarter of the whole universe, to be precise 27 per cent of its mass, is made up of this mysterious, invisible substance.

Studies tell us that dark matter is distributed throughout the universe in the form of an enormous, irregularly woven web; its nodes are concentrated in areas occupied by galaxy clusters, with fine threads connecting the regions of highest density. Enormous spheres of dark matter surround every galaxy. Even our own Milky Way is held together by this mysterious form of matter, which occupies an enormous volume, much greater than that where the stars are concentrated. If the luminous matter of the Milky Way is enclosed within a sphere of diameter little greater than a hundred thousand light years, the dark matter in our galaxy occupies a region ten times as large: a gigantic sphere a million light years in diameter. A similar structure is repeated, on a much larger scale, for cluster galaxies. When we consider the distribution of ordinary matter and of the immense, uneven mesh of dark matter, the universe appears to be a gigantic Gruyère cheese, with large empty spaces separating the clumps where matter is concentrated.

But what is dark matter made up of? This matter, which is so abundant that it is found everywhere, even in the rooms in our houses? Since it neither emits nor absorbs light, it's difficult to detect its presence. It interacts solely by means of the weakest of the forces: gravity certainly and, according to some theories, the weak force too. However, these are forces that are too weak to make it easily possible to track it down. Scientists have developed brilliant instruments to record signals that might allow us to understand its properties. They have sent special satellites into orbit and installed the most sensitive of detectors in underground caverns or abandoned mines under kilometres of rock, as far as possible from any disturbance. But so far, no significant results have been achieved and the hunt continues.

Particles of dark matter are also being sought in accelerators like the LHC. They could be ultra-light particles with strange properties, which are habitually produced but which have escaped us thus far just because we haven't paid sufficient attention in looking for them. Or they could be very heavy particles, which are difficult to produce directly, because the energy of current accelerators is inadequate. If the Higgs decayed into dark matter, a possible albeit rare event, we might spot it if we saw too many bosons disappearing into thin air. The particles of dark matter would pass through our detectors undisturbed, but we would become aware of their presence because we would notice too many events that were unbalanced in terms of energy and momentum.

And finally, there's still the possibility of a great stroke of good fortune, the equivalent of a multi-million-pound lottery win. Candidates for dark matter are present in many theories in the new physics. For example, if a family of supersymmetric particles appeared in the LHC data, we could really take the world by storm! The discovery of supermatter would require us to more than double the catalogue of elementary particles and at the same time we would resolve the mystery of dark matter. We would have made two discoveries in one. But none of this has happened so far.

We must persist, without dreaming too much, because the surprises that await us are not finished yet. Everything we have discussed thus far, ordinary matter and dark matter, only contributes around one third of the total mass of the universe. The best is yet to come.

The dark kingdom of shadows

The discovery of dark energy was a real surprise for everyone, including those working on it. When it happened, in 1998, the astronomers who were the first to find themselves in the presence of such surprising data, couldn't believe their eyes. And yet the results left no doubt. The velocity at which the universe had expanded was not constant; on the contrary, for quite some time now it had been increasing significantly. Everything was moving away from everything at an increasingly frenetic rhythm. What scientists were seeing contradicted what they were expecting; the idea of the accelerated expansion of the universe was counterintuitive. Everyone expected that the attraction exerted by gravity would slowly reduce the expansion velocity of space-time, whereas the exact opposite was happening.

For many years, different teams of scientists tried to understand whether what the data was pointing to was real or whether, on the other hand, errors had been made in the measurements. In the end, they gave in to the evidence. There was no doubt that a new natural phenomenon was being observed, however completely unexpected it was. In the end even the Royal Swedish Academy of Sciences in Stockholm recognized the importance of the work of Saul Perlmutter, Brian Schmidt and Adam Riess, the three astronomers who had carried out the early research, rewarding their discovery with the 2011 Nobel Prize.

Right from the earliest moments, in an attempt to explain this strange phenomenon, the expression *dark energy* was coined, indicating the complete ignorance of the mechanism that produced it: an absolutely unknown form of energy seemingly pushing everything away from everything else and growing as the dimensions of the universe grow. Some imagined a kind of anti-gravity, an extremely strange behaviour of gravity which from being attractive, as we know it, becomes repulsive over great distances. Others imagined a kind of vacuum energy, a positive energy, which creates a kind of negative pressure, thereby pushing everything towards dilation.

The idea that the void contains positive energy which makes it expand goes back many years. And Albert Einstein was the first to come up with it. To make the universe static, that is to counterbalance the effect of gravity, which, acting alone, would sooner or later make everything collapse into one point, Einstein added a positive constant, called the *cosmological constant* into his equations by hand, that is to say arbitrarily. This classification served to build a balance; making the universe expand countered the effects of gravity and made it stable.

Later, when it was discovered that everything had had a turbulent beginning and that galaxies were still moving apart from one another, Einstein regretted this choice, to the extent of referring to it as one of the worst blunders of his life. In fact, with a universe arising from an ultra-dense and super-incandescent singularity, there was no need for this further impetus to expansion to produce a condition of equilibrium. The curious thing is that nobody, least of all Einstein, could predict that by the end of the twentieth century, the discoveries made by Perlmutter, Schmidt and Riess would bring his cosmological constant back into vogue. And so, it seems as if nature will always end up proving Einstein right, even when the great scientist is convinced that he's clearly wrong.

In this case, too, precious information about the presence and distribution of dark energy can be extracted by analysing the tiniest inhomogeneity in cosmic background radiation and the gravitational lens effects produced by galaxies and clusters. It's curious to discover that it is still light which allows us to take a look at this shady side of the cosmos.

The distribution of dark energy in the cosmos is very homogeneous. It behaves quite differently from matter, whether ordinary matter or dark matter. These latter material substances have reticular distributions with high-density nodes and filaments alternating with broad empty spaces. On the contrary, dark energy is distributed uniformly throughout space and seems to occupy the entire volume of the universe quite happily, exerting a repulsive force on everything.

In an attempt to understand the origin of this mysterious form of energy, scientists have ascertained whether the expansion velocity is the same, over a given period, for all the different regions of the universe. They also realized that this phenomenon has only become dominant in the last billions of years. For a long period, the universe expanded following a very different rhythm from the current one.

Various hypotheses have been tested, including the idea that we are dealing with a new fundamental force or an anomalous behaviour of gravity or even the presence in the fabric of space-time of very particular structures, similar to defects in its regular pattern. But, as yet, nobody has managed to understand what gives rise to this strange phenomenon, and explaining dark energy remains one of the most formidable challenges of modern science.

While the mystery surrounding its origins remains, the precise measurements taken of the effects of dark energy on the geometry of the universe and on the spatial fluctuations in the density of matter have made it possible to quantify the weight of this component in the material composition of the universe.

The result is sensational; dark energy contributes around 68 per cent of the total mass. Around two-thirds of the universe is made up of this most mysterious of components. Totalling up the contribution of dark energy, we obtain a frankly embarrassing result. Despite the great progress made by contemporary science, we are forced to admit that we don't know anything about 95 per cent of everything that surrounds us.

Delicate and gentle messengers

Taking into consideration the contribution made by dark energy, at this point we have reached almost 100 per cent of the total. That leaves only a small percentage, which is important not so much for its, negligible, quantitative value, but because it is rich in information. Maybe some of the mysteries we have been discussing will be resolved when we are in a position to better understand what these mysterious messengers are telling us.

We are talking about neutrinos, those particles, similar to electrons, but much lighter and above all neutral, that we have already encountered in the world of sub-atomic dimensions. But neutrinos find themselves at ease even at the other extreme, in the realm of vast cosmic distances. They roam around between the galaxies in swarms, are continually being emitted by the Sun and the other stars, and they have fun passing through the most massive planets without leaving a trace.

All the principal nuclear fusion reactions which take place in the core of stars produce neutrinos. Every star that lights up the sky emits great quantities of them. Thanks to their extremely weak interaction with matter, they manage to emerge immediately from even the most massive stars, to wander perpetually around the whole cosmos. Not even neutron stars, the densest objects that we can imagine, manage to imprison them. Neutrinos remain caught in the ultra-dense mass of nuclear matter just for a few seconds, then they free themselves from the embrace and fly away.

Enormous flows of neutrinos are emitted even when a star dies. Supernova explosions are characterized by a typical flash of light, but, in fact, almost the entirety of the energy of the gravitational collapse is carried by neutrinos. Each time a large star ends its cycle of life, these silent messengers travel across the entire universe. It will take time, because they won't be able to exceed the speed of light anyway, but, sooner or later, they will succeed in carrying the news of the catastrophe to every corner of the universe.

There are also ultra-high energy neutrinos, genuine cosmic messengers which allow us to investigate the most extreme environments of the universe. By comparison with cosmic rays, neutrinos are insensitive to magnetic fields and travel in straight lines. Only weakly interacting with matter, they are indifferent to everything they pass through and can calmly travel from one galaxy to another. If we piece together their direction of flight, we can find out the position of the source that emitted them.

Cosmic rays, on the other hand, are made up of charged particles, principally protons. Chasing around the galaxy where they were generated, they interact with everything they encounter: magnetic fields, dust, gases. For those with the highest energy, even the photons from cosmic background radiation represent a target they can't ignore. And so, these are not the most useful tools for understanding where they came from and thus the process that generated them.

For this purpose, on the other hand, gamma rays, bursts of high energy photons, are useful, because they too can travel almost undisturbed through the vast intergalactic spaces. It is precisely thanks to the coincidence between a high energy neutrino and a gamma burst that there was success, recently, in getting some initial indications about the origin of some of the most violent phenomena in the cosmos.

To identify high energy neutrinos, enormous volumes of water or ice have to be set up and thoroughly shielded from the flow of cosmic rays that would affect the surface. The detector which initially discovered this new class of events is located in Antarctica and is known evocatively as *Ice Cube.* Behind its light-hearted name hides a gigantic piece of apparatus, which exploits the incredible transparency of the Antarctic ice. The Ice Cube researchers have installed light sensors in a one cubic kilometre 'ice cube', around 1,500 metres below the surface.

In those dark and silent depths, signals from highly energetic neutrinos were recorded for the first time. Some of these, interacting with the ice, released charged particles which produced a trail of luminous signals. The energy released in the interaction proved to be sufficiently high to trigger various sensors positioned in the ice, thereby allowing the scientists to piece together the direction of flight of the incident neutrino.

Ice Cube has revealed neutrinos that have energies thousands of times higher than the ones typical of the LHC: unmistakable evidence that they were produced by mighty cosmic acceleration mechanisms.

One of these events, with energy levels lower than the record, but nonetheless equal to around 300 TeV, was recorded in 2017 and immediately sounded the alarm. Knowing the galaxy that the neutrinos were coming from, other pieces of apparatus, specifically those equipped with instruments more sensitive to gamma bursts, were able to verify that ultra-high energy photons had been emitted from the same source. The coincidence of these two phenomena cannot be random, not least because the galaxy responsible contains a black hole which continues to put on a display and to radiate in many frequencies.

Since high-energy neutrinos and gamma rays are the offspring of ultra-energetic cosmic rays, by using this directional information it is possible to identify the most powerful particle acceleration mechanisms in the cosmos: those which, in all probability, are also the source of ultra-high energy cosmic rays.

The most obvious candidates are explosions of supernovas and the activity of supermassive black holes at the centre of galaxies. It is thought that cosmic rays with energies of up to 1000 TeV are being accelerated in explosions of supernovas in our galaxy. The very strong shock wave that spreads through interstellar gas is capable of accelerating particles and nuclei at very high energies. To reach even higher ones, we need to turn to even more terrifying phenomena, such as the paroxysmal phases of the most gigantic black holes, active galactic nuclei, which emit enormous quantities of matter through their relativistic jets. These, in all probability, are the cosmic accelerators whose performance makes a mockery of the LHC, the terrestrial accelerator that we are so proud of.

We find a sea of neutrinos at the other extreme of energy too. Nobody has yet discovered them, but scientists are all in agreement about the existence of a population of fossil neutrinos of cosmological origin drifting around the universe alongside cosmic background radiation.

Cosmological neutrinos are a form of dark matter, but they behave very strangely. Both ordinary matter and dark matter come together to form structures, something which doesn't happen with neutrinos. This ultra-cold matter has an average temperature of 1.95 degrees above absolute zero, even colder, in fact, than the photons of the cosmic background radiation.

But that minimum quantity of energy endows cosmological neutrinos with a velocity that is high enough to allow them to chase around all over the place.

In the primordial universe, the same low-energy neutrinos which make up the cosmic background flew around at ultra-relativistic velocities, and their role was very important. Today, their function is certainly reduced but they remain a fundamental source of information for our understanding of many details about the infant universe. Their separation from matter occurred merely one second after the Big Bang, whereas photons took hundreds of thousands of years to succeed in freeing themselves from its embrace. This means that imprints, characteristic of the primordial universe, have remained in the properties of the cosmic neutrino background, which, if correctly interpreted, could yet reveal many more secrets. The hunt for the first signals from cosmological neutrinos is still on.

And so, neutrinos are an abundant species in the universe, even if their contribution to the total mass amounts to a negligible 0.3 per cent.

These delicate, gentle messengers have good manners, they don't disturb or bother us, but if we listen attentively to what they have to say, they bring us dreadful news of worlds visited by unimaginable catastrophes or tell us about the white-hot environment of our universe's first moments of life. And so, it's the most polite and self-effacing of particles that undermine many of our assumptions, and whisper into our ears the most disquieting and frightening of stories.

8

What supports the water in which Bahamut swims?

Bahamut is a colossal fish, which features in several Arabic legends about the origin of the world. It is quoted in one of the stories of *One Thousand and One Nights* and didn't escape the attention of Jorge Luis Borges.

In Borges' *Book of Imaginary Beings*, Bahamut is referenced thus:

> God made the earth, but the earth had no base and so under the earth he made an angel. But the angel had no base and so under the angel's feet he made a crag of ruby. But the crag had no base and so under the crag he made a bull endowed with four thousand eyes, ears, nostrils, mouths, tongues and feet. But the bull had no base and so under the bull he made a fish named Bahamut, and under the fish he put water, and under the water he put darkness, and beyond this men's knowledge does not reach.*

The legend of Bahamut is reminiscent of the situation that late twentieth-century cosmology has had to come to terms

---

* Quotation from Jorge Luis Borges, *Book of Imaginary Beings*, translated by Norman Thomas di Giovanni, Penguin, 1974.

with: to find, for every cause which produces a certain phenomenon, an antecedent cause which triggers that mechanism, and so on to infinity. It is an infernal trap into which we risk falling every time we try to narrate the birth of matter. 'OK, so there was the Big Bang, but what caused the Big Bang, then?' But it's a trap that we are now able to avoid. To paraphrase Borges' words, at the end of the twentieth century human knowledge has penetrated the darkness which supports the water in which Bahamut swims.

The critical density of the universe

The Big Bang theory, right from its earliest formulation, provoked extremely heated debate. The assumption that the universe was eternal and immutable was very widespread even amongst the most eminent scientists. To counter what seemed to be an overly suggestive hypothesis, they had set the steady state theory against it. The alternative cosmological hypothesis to the Big Bang accepted the idea of an expanding universe but hypothesized a continuous creation of new matter inside it, keeping its density constant. In this way, the universe would, over time, maintain its properties, without having to imagine a beginning or an end to everything. The mechanism was a little over-complex, but the creation of new matter would have been difficult to prove, because it was a question of giving birth to just one hydrogen atom per cubic kilometre per year: a quantity of matter so negligible that falsifying the theory would have been impossible.

The two rival cosmological theories fired up scientists' minds for several decades, and one of the most stubborn supporters of the steady state theory was that same Fred Hoyle to whom, ironically, we owe the coining of the expression Big Bang.

Hoyle was a fervent proponent of the materialist hypothesis, fiercely opposed to the idea of an initial singularity, precisely because it resembled too closely the idea of creation upheld by most religions. Even more irksome to him, without a doubt, was the fact that the father of the theory that upset him so much was Georges Lemaître, a great physicist, but also a man of faith and a Catholic to boot.

The discovery of cosmic background radiation put an end to all discussion, even if, like the last Japanese fighters found in the jungle decades after the end of the Second World War, Hoyle and others obstinately continued to defend their theses until his death in 2001.

In any case, the bulk of the scientific community did not follow him on this kind of personal crusade. Penzias and Wilson had discovered exactly what those on the side of the Big Bang had predicted. And the empirical data on the uniformity and temperature of this very characteristic radiation corresponded to the data that had been calculated on the basis of theoretical models. At this point there were no more doubts; the universe was born billions of years before, from a very special point in which density and temperature reached ultra-high values.

But this is where Bahamut's trap kicks in. If everything was generated in the Big Bang, what gave rise to the Big Bang?

The question was very pertinent, because it wasn't sufficient to talk about a singularity; the most demanding, even among the scientists who were supporting this new theory, raised the problem of the dynamics. No known mechanism was able to concentrate into one point of infinitesimal dimensions, everything that is contained in the universe that surrounds us. The paradox was obvious. Physicists, accustomed to explaining the dynamics of all natural phenomena, failed to offer any explanation as to the mechanisms that would have been able to trigger the mother of all physical processes, the original production of matter. The question then assumed an embarrassing profile if looked at from the point of view of energy. If, to produce the gigantic universe that we know, it was necessary to expend an enormous quantity of energy, some people were asking themselves: so, who or what generated that energy?

For many decades, while the evidence in support of the Big Bang theory continued to build and to reinforce the cosmological hypothesis which by now was considered unrivalled, this intrinsic weakness remained. Nobody was able to answer the simplest of questions: what produced the Big Bang?

And so, people started to think that the answer might come from looking at the future of the universe. By trying to understand its evolution, and maybe its end, they might chance upon the right solution to understanding its beginning. So, the question became interconnected with the necessity of measuring the density of the universe.

In the hypothesis that the universe is expanding, the density of its content of mass and energy is extremely relevant, because the gravitational attraction that is opposed to the increase in volume is dependent on this. The greater the universe's content of mass and energy, the greater the impulse to oppose its expansion, to the point, eventually, of reversing its course.

In its turn, the density of the universe had a relationship with space-time geometry. If gravity dominates, the universe is closed, spherical, finite and destined to collapse in on itself. Space-time is curved and gravitational attraction slows down the impulse towards expansion, succeeds in stopping it and ends up reversing it. Everything would start getting closer and closer together until it collapses into a point of almost infinite density. The singularity would be recreated. The universe would pass through an infinite cycle of phases of expansion, starting with the Big Bang, alternating with phases of contraction, which lead to a Big Crunch, the great gravitational collapse.

This cyclical mechanism would allow us to understand the dynamics of the whole process and would avoid, in principle, the trap of wondering what triggered it. But if this were not the case, the problem would re-emerge in its entirety. If the impulse to expansion dominates, the universe becomes infinite and destined to unlimited growth. Space-time would prove to be curved in the shape of a saddle, and the universe would have a hyperbolic geometry. If, on the other hand, the two impulses were equivalent to each other, we would have a flat universe, which follows Euclidean geometry and would be destined to eternal expansion.

Measuring the density of mass and energy of the universe therefore became essential in understanding its evolution and trying to see whether the trap of the dynamics of the Big Bang could be avoided.

All the observations carried out with the most varied instruments – satellites specializing in precision measurements of cosmic background radiation, analysis of the abundance of primordial elements, use of gravitational lensing to assess the distribution of mass in the universe – all provided unambiguous results. The estimate of critical density was five nuclei of hydrogen per cubic metre, while the average density of ordinary matter was 0.25 nuclei per cubic metre. The density of the universe was twenty times lower than the critical density, and therefore the universe could not be closed. The hypothesis that everything could be explained by alternating Big Bangs and Big Crunches had to be abandoned forever.

Thus, the need to offer a dynamic explanation for the initial singularity re-emerged. The theory continued to elicit consensus, but the most astute scientists were aware that if a convincing explanation for the mechanism that had triggered the Big Bang were not found, the whole construct could collapse in an instant.

At this point, a new fact intervened, a revolutionary theory, developed by a couple of young physicists in the early 1980s. The hypothesis that it was cosmic inflation which had triggered the Big Bang was advanced, in different versions, by the American Alan Guth and the Russian Andrej Linde. The idea was attractive because it placed a random mechanism at the origin of everything, a quantum fluctuation which extracted from the void a scalar particle, the *inflaton*. This newcomer could have sparked that terrifying exponential expansion which we have called the Big Bang.

The theory sparked much debate, but initially not many people took it seriously because it seemed to contradict the experimental data. Things changed when precision measurements of cosmic background radiation started to pour in.

The inflationary model was capable of explaining the great homogeneity of temperature recorded by the most sensitive apparatus. Only exponential growth which had expanded the most minute fluctuations onto a cosmic scale could explain how it was that regions of the universe, billions of light years apart, were exactly the same temperature.

But there remained a glaring contradiction. The inflationary model requires a universe with flat curvature, Euclidean geometry and density equal to critical density. And this is where all the data seemed to prove the proponents of inflation wrong. It was only with the discovery of dark energy and the correct assessment of the contribution of dark energy and matter to the total mass of the universe that the pieces of the puzzle fell into place. A detailed analysis of the distribution of the smallest anomalies, the so-called temperature distribution anisotropies, brought further elements which confirmed the cosmic inflation theory. The new theory had demonstrated a predictive power which was then confirmed by experimental data.

What the vacuum is

All the measurements carried out up until today favour the inflationary model. If not all critics are yet convinced of its validity, it's because we still haven't tracked down the inflaton, the scalar particle responsible for this paroxysmal swelling, the particle which triggered the initial Big Bang. That a certain scepticism should persist, therefore, among some in the scientific community is normal, a part of the usual dynamics. Until the smoking gun of cosmic inflation, the particle responsible for this row, is found, it will not be possible to access the many details which still remain to be defined. However, the general picture is already sufficiently accurate and turns out to be quite surprising.

To better understand what matter emerged from, what the origin of the universe is, it's helpful to return to Fred Hoyle's objections to the Big Bang theory. The British physicist saw it as mere smoke and mirrors, because he was unable to accept a theory which violated Lavoisier's principle. The great French chemist, who would end up being guillotined during the Revolution, had formulated his law of conservation of mass thus: 'In a chemical reaction which takes place in a sealed container, the sum of the mass of the reagents is always equal to the sum of the mass of the products of the reaction.'

To counter the Big Bang theory, Hoyle returned more or less to Lavoisier's argument: 'But isn't the idea that a lot of stuff, the entire universe, was born in an instant, from nothing, even more paradoxical?' His statement echoes the words of the great Greek thinker, Empedocles, who lived in Agrigento in the fifth century BCE: 'Nothing comes from nothing.'

And this is the crux of the question. What gives birth to the universe produced by the Big Bang? Contemporary physics replies: the void, otherwise known as the vacuum. But beware; the vacuum is not nothingness, on the contrary, in certain respects it's the opposite of nothingness. And so, to answer Hoyle's or, if you prefer, Empedocles' objections, we are obliged to fully understand what the vacuum is.

Let's take a cubic metre of air in front of us; let's imagine a large box, one metre by one metre, right in front of us, in our office or our dining room, and let's imagine wanting to transform it into a cubic metre of vacuum. What procedures do we need to carry out?

First of all, we need to remove all the air. Before doing this, we would have to get hold of a powerful suction pump and enclose this volume in a strong container, capable of coping with the pressure, because, if we suck out the molecules of air from the inside, the air outside the cube will exert pressure on its walls. Not a problem: get a nice thick steel cube and we can proceed. As a vacuum is gradually created inside the cube, we will have to make use of ever more powerful pumps. The technology that we have at our disposal today does not allow us to extract all the molecules right down to the very last one; there would always still be a residual pressure from the gas which, however minimal, would correspond to a very large number of molecules still present in the cube. But let's suppose we've found a system by means of which all the molecules, one after the other, were extracted from the container. Now, inside our cube, there is a vacuum that is more extreme than the intergalactic one.

Once the air has been taken out, what's left inside the cube? So many things. Above all there are probably electromagnetic fields. It's true that the steel forms a kind of shield, but certain frequencies of the tide of electromagnetic waves in which we are immersed, could, nonetheless, penetrate that shield. To avoid this, we would have to increase the shield and make sure that it's capable of attenuating all frequencies by many orders of magnitude.

Then we would have to deal with the photons emitted by radiation on the inside. Our container is in equilibrium with the temperature of the room, and therefore its walls, being hot, emit photons. And so, we have to cool everything down. Very complicated but let's suppose we manage to do it, even if the apparatus is now becoming more and more complex. We need to have access to a large system of refrigeration, and we have to isolate the steel cube from everything surrounding it, otherwise it will heat up again. To cool it down such that its walls no longer emit photons, we will have to take it to absolute zero, −273.15 degrees centigrade. This is something we cannot actually do. But let's suppose here, too, that we can find a way of reaching the temperature we had proposed. Now is there a complete vacuum inside? No chance.

The volume occupied by the cube is crossed by cosmic rays which are continuously raining down on us and by the flow of neutrinos arriving from all directions. To shield our cube from the former is perhaps conceivable. We would have to take the cube to a subterranean cavern protected by kilometres of rock to absorb a large part of the cosmic rays raining down on us. But as far as the neutrinos are concerned, the battle is lost before it's begun. Not even the thickness of the Earth would be sufficient to stop this uninterrupted flow of such light particles. However, let's imagine even here that we've found the solution that, up till now, no scientist has ever managed to find. Unfortunately, that's not the end of the story.

The vacuum that we have produced would still be *full* of material substances. Most especially the gravitational field. The space-time enclosed by the cube is still curved because of the Earth. If a tiny grain of dust were to detach itself from inside the top of the cube, it would fall to the bottom as a result of the force of gravity. And then there's the Higgs' scalar field, the one that gives mass to all elementary particles, which remains in the cube, indifferent to and unperturbed by all our activity. These are all things we cannot shield against. But even in this case, let's suppose that we've found a way of doing it.

The hardest part would still await us: the one where we are wandering around in complete darkness. In the volume there would still be dark matter, those unknown particles which are all around us and, above all, the space occupied by the cube would still contain dark energy. Not having the faintest idea about the origin of these two material components of the universe, we can't even imagine a way of being able to remove them. But nothing stops us dreaming that it might be possible.

Inside the cube now there is a vacuum, but the material void we have achieved is not nothingness. On the contrary, it contains a myriad of things. It's full of hidden, silent activities.

The vacuum state of a material system is defined as having minimal, ideally zero, energy. But saying that inside the vacuum there is no energy is not the same as saying that inside the vacuum there is nothing. Even the vacuum, like all material states, must obey the laws of quantum mechanics and, in particular, the uncertainty principle. The vacuum is prohibited from having a set of microscopic states *permanently* at zero energy.

The vacuum is full of fluctuations; for one instant, a length of time compatible with Heisenberg's principle, a particle-antiparticle pair can appear in it, and then be reabsorbed immediately afterwards. On the microscopic level, the vacuum is bubbling away, continuously producing a kind of fine foam, a frenzied saraband of tiny bubbles, minute portions of space-time, in which, for an infinitesimal moment, pairs of electrons and positrons or other components of matter and anti-matter survive. The average energy will be zero, but the zero value will arise from an infinite series of states which deviate from the average value. Thanks to this mechanism, an infinitesimal fluctuation of the vacuum can fill up with inflatons and give birth to a material universe.

Where matter comes from

The inflationary model envisages a universe born by chance, from a quantum fluctuation of the vacuum. One of the most convincing proofs in support of this theory is provided by precision measurements of the space-time geometry and of the critical density of the universe.

All the data gathered by the most highly evolved measuring devices agree, with a minimal margin for error, about the fact that our universe is flat, with practically zero intrinsic curvature. Space-time is only deformed locally as a result of the presence of massive celestial bodies. The geometry of the universe is Euclidean, the one that is studied at school; if you draw a triangle by connecting three dots, the sum of the internal angles is exactly 180 degrees. This result, which is anything but taken for granted, arises from the condition of being seated astride the critical density.

It's important to emphasize once again that, at the time when the theory was formulated, the debate over the geometry of the universe was still ongoing. The dominant thesis was that the universe was open and its geometry hyperbolic, given that the sum of ordinary matter and dark matter barely reached one third of the critical density.

It was only the wholly unexpected discovery of dark energy which dissolved many of the reservations over the credibility of cosmic inflation. For about twenty years now, we have been able to explain the origin of the universe in a wholly new and, in certain respects, surprising way.

Saying that the universe is in a condition of critical density is equivalent to saying that its total energy is zero. And this result seems mind-boggling. What? If a drop of water, transformed into energy according to Einstein's mass-energy equivalence relation, corresponds to an enormous value, let's imagine what would happen if we transformed the hundreds of billions of stars in every galaxy into energy and then multiplied the whole by the hundreds of billions of galaxies, finally adding in the contribution of gases, dust, dark matter and dark energy. The energy of the gigantic universe in which we live must have an enormous value which can only be expressed with numbers to the power of ten with a very large exponent. All true, but we always overlook one piece of data. In the universe there is not only mass-energy, but also another fundamental component, space-time. Our universe is made up of these two ingredients combined together. A great quantity of mass-energy distributed across an enormous structure which we call space-time.

But beware, this structure is not a non-material structure, or an abstract concept: quite the opposite. Let's take the space-time which surrounds us; it contains fields, like the electromagnetic one which allows us to communicate, or the gravitational one which keeps us attached to the ground. Space-time, despite being extremely rigid, is a material substance that can vibrate, oscillate and transmit energy across great distances through gravitational waves.

If we consider this from a more general point of view, we realize that space-time plays a decisive role in the energetic balance of the entire universe. Since the warping of space-time due to the local presence of a concentration of mass-energy produces gravitational attraction between two bodies, so we find ourselves having to imagine it as if impregnated by a particular form of energy, the binding energy which holds together two bodies which are attracted to each other.

We have already seen that to free a body from the embrace of the Earth, we would need to supply it with a velocity of 11 km/s. The kinetic energy relative to the velocity of the body that we want to send into outer space precisely corresponds to its binding energy, which is always negative; by adding them to each other, it turns out that the body has zero total energy, and thus can travel to the point where it will no longer feel Earth's pull and which it will reach at a velocity that will reduce slowly until it disappears altogether.

Every pair of celestial bodies scattered around the universe has a corresponding binding energy. Since everything is attracted by everything, the sum of energy that keeps the stars, the galaxies and the clusters of galaxies together is equivalent to an enormous negative number. And then there is dust, gas, dark matter and the negative number becomes even bigger. If we subtract the two gigantic numbers, one positive and one negative, which we have just acquired, one from the other, we get a surprising result: the total energy of the universe is zero.

The positive energy of mass-energy and the negative energy of the gravitational field, contained in space-time, cancel each other out if the universe is flat. This means that the density of the universe is equal to the critical value. A universe of this kind can expand to infinity and last for eternity because its dynamic does not require any expenditure of energy. It's like a tight-rope walker moving, with extreme lightness and agility, along a high wire. It is obvious that all this cannot be random. Beneath this apparent paradox a deep truth is hiding.

The energy of the universe is zero, just as its total charge or its overall angular momentum is zero; for every group of galaxies orbiting around in a particular direction, an equivalent one has been found whose components are rotating in the opposite direction. And so, the physical properties of the universe system, what we call quantum numbers, are all worth zero, exactly like those of the vacuum system. The conclusion is inevitable; the universe is also a vacuum system. The two are indistinguishable from a physical point of view.

Accepting this conclusion proves rather difficult for human beings used to living in a system which appears to be *full* of material objects: our house, mountains, the Earth, the Sun, the stars and other galaxies. Everything, including our bodies, seems to say to us: the universe is full of material structures, it is itself a gigantic material structure. And instead, contemporary science tells us: beware, this is an illusion, one of the many assumptions which were our constant companions until we got to the bottom of these questions.

If we consider this whole affair from a more rigorous point of view, we discover that our universe is still a form of vacuum, and this discovery radically alters the question of its origins. There's no longer any need to seek an immense source of energy if all you need to produce is an object with zero energy. The thing can happen spontaneously, precisely because it doesn't require any expenditure of energy. The Big Bang was a noisy transformation of one state of vacuum into another state of vacuum, equivalent to the first; a great metamorphosis which, not requiring any energy, can happen free of charge and carry on through time over billions of years. Everything happened by chance, while respecting the one restriction; nobody is allowed to violate the laws of physics which regulate the behaviour of material bodies, all material bodies, even the vacuum state.

At this point modern science has avoided Bahamut's trap. Our material universe was born out of the vacuum, thanks to one of the many quantum fluctuations which characterize it; a microscopic bubble which set out on a spectacular path, so extraordinary that for many years it deceived us, pushing us to think in terms of an initial singularity whose dynamics we didn't understand.

Today we know that everything was much simpler. That this mix of mass-energy and space-time can develop naturally, spontaneously, completely randomly, precisely because the two quantities are complementary from an energy point of view. The positive energy needed to create the former from nothing is identical to the negative energy which the latter turns out to be imbued with. By combining space-time and mass-energy, there's no need to ask to borrow energy from anyone.

Maybe some of this had already been intuitively understood by the great Greek thinkers. It's hard to resist the temptation to interpret Hesiod's poetry from today's perspective when he speaks of Chaos, understood as an endless maelstrom, an empty space from which everything originates. Or the words written by Plato in *Timaeus*: 'Anyway, time came into being along with the heavens so that having been generated simultaneously, they would also be dissolved simultaneously if they were ever dissolved.'*

---

* Quotation from Plato, *Timaeus*, translated by David Horan, Platonic Foundation, 2021.

What will happen to matter in the end

The origin of the universe, the first spark of everything, has so far stolen the limelight from its negative complement: the end. But what will the final act of this glittering spectacle of stars and planets, galaxies and neutron stars, pulsars and black holes be like?

The discussion about the end of the material universe which surrounds us is just as fascinating as the one about its beginning, even if slightly more concerning. Let's start by excluding some of the hypotheses that were highly fashionable until a few decades ago. We have already hinted at the fact that the hypothesis of a cyclical universe, an alternation of Big Bangs and Big Crunches, has already been abandoned. Since the discovery of dark energy, we know that there will be no great contraction in our future.

Today, there are no longer any doubts; the universe is expanding at an increasing velocity. Of course, some galaxies will collide with others and maybe fuse, as is likely to happen between our Milky Way and Andromeda. But the universe seems to be resolutely pushing towards rarefaction. Although it is difficult to make predictions on a timescale of tens of billions of years, if this phenomenon persists, everything will move away from everything else.

In an ultra-rarefied cosmos, distances will become too great to make the formation of new luminous stars possible. When one generation of stars reaches the end of its life cycle, no others will be born to replace them and the whole universe will become an infinite expanse of dead stars: brown dwarfs, neutron stars and black holes which will wander around indefinitely in the dark and the cold. This is what we call *heat death*, similar to a suspended life, which the English have nicknamed the *Big Freeze*.

The universe and its matter, of course, will continue to live for billions of years. It's just that, inside that dark, cold universe, the conditions will not exist to create the dynamics that we are connected to, made up of stable solar systems, warm inhabitable planets and sufficient energy to build and develop material forms as complex as biological ones.

As far as dark energy is concerned, we know its effects, but being wholly unaware of its origin and nature, we are not able to predict its evolution over extended timescales. All we can do is hypothesize. If, for example, its acceleration should increase exponentially, we might even end up with a laceration in the fabric of space-time. Even today, it's hard to imagine it as something material, that can be torn apart, but the possibility should not be excluded.

Space-time is an extremely rigid structure, but it's not indestructible. A great rip would have catastrophic global effects; we only need imagine what would happen to the gravitational bonds between stars and between galaxies. We don't know what would happen to the other interactions which hold matter together on the microscopic level. Some suggest that the laceration would end up smashing all material forms to pieces; even protons and neutrons would vanish, not to mention atoms and molecules. The universe would go back to being an immense structure in which elementary particles drift, totally isolated one from the other and incapable of joining together. A structure similar to that of the first few moments, but infinitely larger, older, colder.

In the scenario of the heat death of the universe, some suggest the possibility of the creation of black holes which are so gigantic as to produce a laceration of space-time. But even here we are in the realm of speculation that is difficult to prove.

As an alternative to this sad and rather depressing scenario, we should bear in mind another recent hypothesis concerning the end of the universe: the so-called *electroweak vacuum catastrophe.*

With the discovery of the Higgs boson, questions arose around the stability of this mechanism, which is so important in giving mass to elementary particles and allowing matter to organize itself into permanent forms. As soon as we succeeded in measuring the mass of the new particle, it was possible to carry out a study of the properties of this very particular field and there was no shortage of surprises, the most relevant of which was the discovery that it is in a state of metastable equilibrium. Described thus, it would seem of little interest, even irrelevant. In fact, though, what was discovered is very intriguing and conceals worrying aspects. The strange potential linked to the Higgs boson, in the very first moments of the life of the universe, is located in a particular condition of equilibrium in which it has been resting for 13.8 billion years. The question we asked ourselves is whether this position of equilibrium is absolutely stable or not. And the answer turned out to be different from what we were expecting. Given that the system has existed for billions of years, we were expecting to find conditions of absolute stability, that is to say that this delicate equilibrium was constructed so well as to be able to sleep soundly for many more billions of years. In fact, that's not how it is. The equilibrium is actually metastable, which means that something could disturb it or break it. And so, it is not at all guaranteed that what has happened up till now will continue to happen forever.

In other words, the kind of delicate scaffolding which gives mass to elementary particles could suddenly collapse. If this were to occur, there would be trouble; all bonds would break and organized matter – stars, planets, basically everything, including us human beings – would disintegrate in the moment, transforming into pure energy. Some people hypothesize an enormous bubble expanding at the speed of light; for others the catastrophe would affect the whole universe straightaway. The only consolation, if it can be called that, is that it would in any case be a very hot ending, and very spectacular, in certain respects the opposite of that depressing condemnation to eternal cold and darkness implied by heat death. But beware, the universe would continue to exist for billions of years, in the form of rarefied particle plasma, but it would be wholly different from the one that is currently our home.

In what conditions could this catastrophe occur? Studies tell us that the electroweak vacuum could collapse under the pressure of a large quantity of energy, levels that we cannot, even remotely, imagine reaching on planet Earth. And this is a good thing because I'm certain that, otherwise, someone would come up with the idea of building an Armageddon bomb for us. But the fact that these levels of energy are unattainable for us should not offer too much comfort. We have, on too many occasions, been surprised by wholly unexpected natural phenomena. The collision of two black holes which made possible the revelation of gravitational waves is one of these. In this specific case, nobody can exclude that in some hidden corner of a distant galaxy, mechanisms as yet unknown to us, might be able to trigger levels of energy high enough to shatter or fuse the electroweak vacuum. By suddenly shifting it from that serene position of equilibrium, which it has occupied for 13.8 billion years, the vacuum would go back to being simply a vacuum, incapable of carrying out its essential function in the construction of those material forms we are so fond of, including ourselves.

And so, whether we imagine heat death in the various scenarios previously described, or whether we consider the electroweak vacuum catastrophe, for us humans, assuming that any of us would be around when these phenomena occurred, things would not be good at all.

In the last fifty years, a range of theories have been developed which predict the existence of a vast number of universes, each one different from the others. In this case, we could always console ourselves by saying that nothing too serious would happen if our universe came to an end, given that all the others would continue to exist. I'm afraid that, for the moment, we have to give up this meagre consolation, given that these theories, of the so-called *multiverse*, are just elegant conjectures, that is to say theories that have not been proved experimentally and which are consequently not a sensible basis for rash speculation.

9

The magnificent illusion

His works are immediately recognizable from the materials used and the dramatic roughness which is typical of them. Whether he's using sacking, plastic scorched with a blowtorch or cracked clay, the material itself assumes a central role in all of the work of Alberto Burri, one of the masters of matter painting.

One of the greatest artists of the second half of the twentieth century produced his works by ripping, burning, sometimes reconstituting the different forms of material he was using. Some have seen in those rips the evocation of a wound, the tatters of tormented flesh, humanity's cry of pain or anguish, suddenly finding themselves facing the risk of nuclear holocaust having just gone through the horrors of a war.

This interpretation finds confirmation in Burri's own life experience: a surgeon who took part in the Second World War as a medical officer to the Italian troops in Africa. Burri was a committed fascist, a black shirt, utterly loyal to Mussolini. Like many Italians at the time, he too was certain Il Duce would rebuild the Empire, so much so that as early as 1935, he followed the army of occupation to Ethiopia.

In March 1943, Alberto Burri at the age of twenty-eight is a medical officer, part of the Xth Mussolini Battalion. He's sent to Africa with his company to fight the British, but a few months later, he's taken prisoner. On 8 May, Burri is captured just outside Tunis and sent to a prisoner of war camp in the United States, along with thousands of other Italian soldiers.

At Hereford, near Amarillo, Texas, there are over five thousand Italian prisoners and living conditions are tough, but tolerable. The situation becomes trickier after September 1943, the date of the Armistice. In the Spring of 1944, the Italian prisoners divide into two camps, those who are distancing themselves from Mussolini and support the new Badoglio government, and the so-called hard-core fascists. The former are liberated, and sent back to Italy to fight with the Allied troops. For the others, their detention continues and becomes tougher and tougher. Burri is one of these prisoners, subjected to extremely harsh conditions, with frequent hardship and abuse. It's around this time that he starts to dedicate himself to painting.

Later he will remember his beginnings like this: 'I would paint every day. It was a way of not thinking about everything around me and the war. I did nothing but paint until the Liberation. And those were the years when I realized I "had" to be a painter.'

In fact, with the end of the war and Burri's return to Italy in 1946, he gave up his profession as a doctor to dedicate himself wholly to painting. And he began the search which would lead him from oil colours to the use of paper, cardboard, hemp sacks, tar, wood, plastic and scrap iron. These are only a few of the materials he used, often mixing them with colour. He selects and juxtaposes different materials, tears them, rips their structure to pieces, puts them back together again, burns them. His works are full of welding joints, stitches, twists. As early as the end of the 1940s, in total solitude, Burri is making a radical break with traditional pictorial language. It will take several more decades before the whole world acknowledges that one of the greatest artists of the twentieth century is standing before them.

There's no doubt that the traumatic experience of the war plays a role in his work, but I share the view of those who consider this interpretation to be highly reductive. The works of the master of matter tell us something much deeper.

As often happens, works of art acquire their meaning, in the eyes of the viewer, from the interwoven dialogue between the sensibilities of the artist who created them and the experience and mental structures of the spectator who views them. In the 1970s, when Burri was using a blowtorch to produce burns which tore through plastic reducing it to a mass of holes and congealed filaments, he could not have imagined that decades later, to my eyes, those structures should call to mind the web of dark matter which enfolds the cosmos. I have always seen, in his works, an attempt to give form to what has no form, to depict the effort made by matter to organize itself and acquire meaning. Alongside the existential drama, I have always detected a discourse of much wider significance, which concerns us all: a wound transformed into beauty. In this case, the search for beauty in the depiction of the superhuman effort of matter to organize itself into something radically new and precious for everyone.

The great euphoria of mechanistic materialism

Burri's material images shed light on a century, the twentieth, in which science, with its fast-moving advances, came to occupy centre stage. The influence of scientific discoveries on the principal artists of the twentieth century is well documented. But its impact on the figurative arts is only one aspect of a more general paradigm shift, which influences and involves all human disciplines: from theatre to literature, from psychoanalysis to philosophy.

The century we have behind us is the one in which the long dispute between idealists and materialists, which had gone on for millennia, seems to have been definitively resolved in favour of the latter. The foundations of this success were put in place a century before by the two great thinkers, Charles Darwin and Karl Marx. Both had emphasized the importance of material processes, albeit in very different fields.

Darwin had shed light on the material mechanisms lying behind the evolution of living species. Our own biological nature itself – our bodies, the senses we use, the way we walk, speak, interact with our fellow creatures – arises from a long evolutionary history, in which genetic sources and material conditioning by the environments in which we have lived, are interwoven. Today, we know it was not a linear history, that there were abandoned solutions, blind alleys, inefficiencies and backward steps, but nobody questions the basic mechanism anymore.

Marx links the evolution of thought and the organization of all societies to the material structure with which humans organize themselves to produce the goods we need. Anthropological studies and the in-depth examination of an endless variety of societies of different periods in the past and in different regions of the world have led us to adopt a much less mechanistic and more nuanced attitude. But there remains an underlying truth in the original approach; it is a fact that Marx altered humanity's point of view on its social organizations, and it's no coincidence that this new awareness gave rise to some of the greatest social changes in history.

All of this prepares the ground, but it is the tumultuous development of knowledge which cuts through any resistance. Euphoric with all the progress made by science in every field, the late nineteenth-century materialists are intoxicated by their own success. Out of this arises an acritical exaltation of material data in and of itself; the myth of matter as eternal and infinite is praised to the heavens. There is a celebration of nature which is self-generated and destined to last forever, in which fundamental laws are inscribed which cannot be violated. Man is considered to be just one element of nature like any other, who must make his own choices by simply following the immutable laws which regulate the universe.

In this delirious omnipotence of positivist mechanism, the only knowledge worthy of the name turns out to be scientific knowledge. Art, religion, literature and philosophy are no longer considered to be activities capable of generating authentic knowledge. As disciplines, they are similar to dreams, from which, in the best-case scenario, some form of aesthetic pleasure can be derived. Many think that it's only a matter of time; the moment will soon come when every kind of question, including those which concern human beings and their relationships with their fellow creatures, will be resolved scientifically. It will be science which tells us what is right and what is wrong, what is beautiful and what ugly.

In this babble, we can hear an echo of the famous phrase of the great French physicist and mathematician Pierre-Simon de Laplace, who, in 1812, had written what many consider to be the true manifesto of positivist mechanism:

> We may regard the present state of the universe as the effect of its past and the cause of its future. An intellect which at a certain moment would know all forces that set nature in motion, and all positions of all items of which nature is composed, if this intellect were also vast enough to submit these data to analysis, it would embrace in a single formula the movements of the greatest bodies of the universe and those of the tiniest atom; for such an intellect nothing would be uncertain and the future just like the past could be present before its eyes.*

* Quotation from Pierre-Simon Laplace, *A Philosophical Essay on Probabilities*, translated by F.W. Truscott and F.L. Emory, John Wiley and Sons, 1902.

The materialist-mechanistic conception reduces the world to four entities: matter, force, space and time. The latter two are abstract, imperturbable receptacles, totally independent of events and of physical processes which occur inside them. The universal mathematical laws of mechanics which govern motion are sufficient to offer an explanation of every phenomenon that might occur within those receptacles.

This mechanistic vision of everything had become dominant by the end of the nineteenth century. Such and so extensive is the progress made by science in chemistry, in thermodynamics, in mechanics and in many other disciplines that nothing seems capable of contradicting the idea that the whole world behaves like a massive machine.

This conviction, linked to a boundless faith in the future, will collapse as a result of the events which a few years further on will push Europe over the edge. With the slaughter of the First World War, the euphoria of the Belle Epoque and the hope of a world illuminated by progress and knowledge will vanish overnight. The immense tragedies that will accompany and follow the Great War will undermine forever the idea of progress led by the scientific and rational evolution of humanity.

The curious thing, the true heterogony of ends, is that in the same years that intellectuals and public opinion were celebrating the triumph of materialist mechanism, a group of visionary scientists were laying the foundations of a new concept of matter, so unusual and revolutionary, that it would definitively consign materialist mechanism to the archives. With the birth of relativity and quantum mechanics, the most radical critique of that same determinist-materialist model, which seemed so dominant, is launched right from the heart of the most advanced scientific research. While all this was happening, the great upheavals which had led to the October Revolution in Russia, had elevated materialism to such heights of political triumph as to make it a state philosophy.

State materialism

In the philosophical thinking of Vladimir Ilyich Ulyanov, otherwise known as Lenin, the victory of the Russian Revolution represented the clearest proof of the superiority of materialism over idealism. These two schools of thought had challenged and fought each other for thousands of years, but now nobody could be in any doubt anymore. The Bolsheviks' affirmation, with the establishment of the dictatorship of the proletariat, not only brought about that institutional structure thanks to which the class of the exploited and oppressed exercised power, but it was also the incarnation of the success of a new vision of the world, which placed matter, as was its due, at the heart, categorizing everything else as mere superstition and deceit.

And this is how materialism became the state philosophy in Soviet Russia. It was taught to millions of young people, it was studied in the best universities, spreading throughout every field of activity, from education to art and culture in the widest sense.

When the direction of the party was taken over by Ioseb Besarionis dze Jughashvili, otherwise known as Stalin, the approach, if that were possible, became even more systematic and invasive. He set up a specific section of the Central Committee with responsibility for ideological, cultural and philosophical matters. Not a single literary work, ballet or symphony, cinematographic production or piece of artistic and scientific research was not examined in the greatest detail, to assess its adherence to the party's general philosophical approach. For anyone not aligned, the risks were enormous, including deportation and the firing squad.

The most emblematic representative of this paranoid attention to detail was Andrei Zhdanov, personally appointed by Stalin to be head of the organization responsible for guaranteeing the adherence of Soviet cultural policy to the principles of materialism. Artists, scientists and intellectuals would have to align their work with the party line. Under Zhdanov, the situation assumed a mind-boggling profile. The level of absurdity was reached when even scientific laws were thought to be up for discussion. The primacy of the party in the ideological and cultural field admitted neither limits nor boundaries.

The principal targets of a thorough-going campaign to delegitimize modern science were the Big Bang theory, quantum mechanics and relativity. Around the end of the 1940s, Zhdanov decided to unleash a battle to cleanse science of ideas considered bourgeois, deceitful and harbingers of illusions. Not a single branch of scientific knowledge was spared, the axe of the highest custodian of Soviet orthodoxy fell without pity on the fields of physics and cosmology.

But please note, Zhdanov was not challenging the content of these new disciplines, which were in fact being studied and developed in every university of Soviet Russia. It's no coincidence that the generation of physicists from the Stalin era proved to be particularly rich in excellent scientists: Piotr Kapitsa, Igor Tamm, Lev Landau, Pavel Cherenkov to name just a few of those who will be awarded the Nobel Prize for Physics. But we shouldn't forget Andrei Sakharov and Bruno Pontecorvo, the youngest of the via Panisperna lads, who moved to Russia to work because of his Communist convictions.

Zhdanov's ideological fury more than anything struck the philosophical interpretations of the new disciplines, but in the heat of the debate it ended up inevitably engulfing a relevant part of their scientific content.

In documents produced in that era, we are in 1947, we can read passages like this: 'Contemporary bourgeois science supplies clericalism and fideism with new arguments . . . the Kantian subterfuges of latter-day bourgeois atomic physicists lead them to deductions of the "free will" of the electron, and to attempts to represent matter as only some combination of waves and other such nonsense.'*

The uncertainty principle became the object of violent attacks because it was considered to be the core of the idealistic visions of Bohr and Heisenberg. The idea that there might be: 'limits to the precision of our measurements, which cannot be overcome by any real apparatus: limits which depend on the properties of the object being studied, different from the properties of the material explained by classical mechanics.'†

But the fiercest attack was reserved for the Big Bang. The idea of a universe both complete and yet still expanding was described as a 'cancerous tumour which corrodes modern astronomical theory and is the principal ideological enemy of materialist science [. . .] The theory of general relativity alone is not sufficient to derive an accurate cosmological theory.' The Big Bang was stamped as a pseudo-scientific, idealist theory, because it was suggesting a birth of the universe which was too similar to the one described in the biblical book of *Genesis*. The fact that the origin of the theory was the work of Georges Lemaître, a Jesuit priest, as well as being a physicist and cosmologist, increased the scepticism of the Stalinist establishment. The galaxies' *red shift*, an unequivocal sign of their moving away, would merely prove to be a pretext. 'The reactionary scientists Lemaître, Milne and others . . . [use the red shift] to strengthen religious views about the structure of the universe . . . These falsifiers of science wish to revive the fairy-tale of the origin of the world from nothing.'

It hardly mattered that over those same years the theory was assuming the profile of a thorough-going physical model of the

* Translation by unknown author commissioned for the journal *Political Affairs*.

† Quotations here and below are taken from E.A. Tropp, V.Ya Frenkel and A.D. Chernin, *Alexander A. Friedmann: The Man Who Made the Universe Expand*, translated by Alexander Dron and Michael Burov, Cambridge University Press, 1993.

primordial universe, thanks to George Gamow. Supporting the Big Bang in this instance was a notoriously atheistic scientist who was also of Russian origin to boot. But he was guilty of having fled the Soviet Union to take refuge in the United States, something for which he would not be forgiven. His theory was stamped as non-scientific, because developed by an Americanized apostate.

These were the Cold War years, the period of the toughest confrontation between the two great powers who had emerged victorious from the Second World War. In the Soviet Union people lived with the nightmare scenario of an attack by the United States, at the time the only country with nuclear weapons. The development of the nuclear bomb was top of the list of priorities for Soviet research programmes.

With Zhdanov's own death first and then Stalin's a few years later, criticisms of the best Russian scientists ended up dying down significantly. Many of the most heated polemics of the period were forgotten from the moment that those same scientists, who had an absolute mastery of relativity and quantum mechanics, showed that they were capable of using that mastery extremely well to achieve aims considered priorities by the party. On 29 August 1949, at the Semipalatinsk nuclear test site, the first Soviet plutonium bomb exploded.

Modern materialism

Twentieth-century physics definitively consigns every attempt at rough realism and materialistic mechanism to the archives. The new phenomena which quantum mechanics allows us to explore radically change the coordinates of materialism. Modern science takes the universe in its entirety back to material forms, including space-time, but the conception of matter which it uses is profoundly different from the traditional one.

The idea of space and time which are eternal and immutable, inside which intrinsically enduring material forms circulate, is destroyed from the roots up. Space-time is born hand in hand with mass-energy and both constituents have undergone an evolution characterized by numerous transformations and genuine catastrophes. Minimal differences, apparently insignificant details, have produced effects which are decisive for the whole of history to come.

An inflationary phase just a tad longer, a less heavy Higgs boson, or a stronger reaction between that and electrons or the lightest quarks, a varying difference of mass between proton and neutron and we should have had neither stars, nor planets, let alone anthropomorphic monkeys endowed with self-awareness busy trying to explain the world to themselves.

Our universe is born and develops randomly and chaotically; it finds almost stable states of equilibrium rolling along on the edge of the precipice, it creates an impression of solidity and persistence, and, at the same time, it is home to the most changeable and ephemeral material forms that the human mind can imagine.

Beneath all these processes, principles of physics are at work which prove difficult to understand at root. Sometimes they arise from deep symmetries, sometimes they seem wholly arbitrary, even though nobody doubts their accuracy anymore.

A universe that is born spontaneously from the vacuum or rather, that is still today a state of vacuum which, thanks to wholly random mechanisms, has undergone an incredible metamorphosis, is a concept that would have left Zhdanov and his acolytes wide-eyed, but which also leaves us, modern scientists, gasping for breath.

A universe made up, above all, of dark energy and dark matter, incomprehensible elements which compose the vast majority of the material world that surrounds us, nips any temptation towards gross materialism in the bud.

At the root of those first philosophical-scientific speculations is surely that primordial setback, that sense of fragility and inadequacy, which overcame the first humans, mortal creatures exposed to every risk, when they beheld, in awe, the Sun and Moon, mountains and the starry sky.

For thousands of years, humanity saw itself as the most fragile element in an eternal and immutable natural scenario, which unsurprisingly was soon attributed divine status.

But this setback gave rise to the most beautiful things the human species has ever produced: art and philosophy, science and religion. It triggered the impulse to build immortal works: the tombs of the Pharaohs which rival the mountains or the great religious and philosophical constructions, destined to last for centuries.

Today, we discover that that same fragility, which we were so ashamed of, and which represented a constant source of anxiety, is a feature we have in common with the whole material world. Modern science tells us that all material substances endowed with some kind of consistency, suffer from this intrinsic fragility; there is nothing, even among the most enormous structures that can exclude us from this law, from this kind of original sin. To speak of permanent material forms, in a universe subjected to transformations of these proportions, is enough to make you smile.

There is nothing that is eternal and immutable. Our existence might last nearly a century, the life of a planet or a galaxy extends over billions of years, but nothing can relieve us of this destiny of transience, albeit over completely different timescales.

It's as if modern science is closing a large circle which opened up at its birth, thousands of years ago.

But at the very moment of this discipline's greatest triumph, we shouldn't forget that science has very precise limits. It is anything but omnipotent. It is one of the most sophisticated products of our ability to formulate a vision of the world and it has led us to mind-blowing results, especially since the development of the Galilean method, which allowed us to explore the most obscure corners of nature. There's no debate about the power of the scientific method; it is the secret weapon that allowed this discipline to develop at an incredible speed.

But science cannot do everything. The method that has given us such striking results cannot be applied to every aspect of reality. The more efficient it is in identifying the mechanisms which regulate measurable and reproducible material processes, the more it stutters, or turns out to be wholly impotent, when faced with phenomena lacking in these characteristics. And our own lives, not to mention the whole complex community of human beings, are characterized by precisely that kind of phenomenon. We are thinking, for example, about emotions – fear, love – or ethical issues – what is good and what evil – or aesthetic ones – what is beautiful and what ugly. These are all things that are extremely serious and as real as a plane flying or a rock falling. But science has nothing to contribute to these questions, as a matter of principle, precisely because a human community of thinking beings, free and interacting with each other, cannot be treated as a physical system.

None of these phenomena is repetitive and measurable; reactions which arise from the interaction between the individual and the community cannot be subjected to general laws. The illusion that we can understand the functioning of the human brain using the same instruments that allowed us to understand the function of our other organs faded some time ago. Nobody argues anymore like Pierre Jean Georges Cabanis, an eighteenth-century French doctor, famous for his statement: 'The brain secretes thought, as the liver secretes bile.' Things are much more complicated than this.

The same is all the more valid for legal or social questions, those which concern the best way of organizing the cohabitation of individuals or groups within a society, or political ones, which define the values and objectives of a community. When there is no standard, when repeat experiments are not possible, we scientists need to keep out of the way. Science cannot pronounce on ethical or aesthetic questions or on the best way of organizing a society or of judging a work of art. And thank goodness that is the case. Beware of giving science a task that lies outside its competence. Scientists cannot be tasked with dealing with more than they are capable of.

But what is matter, really?

Even if we limit ourselves to the world of reproducible phenomena, the modern conception of matter conceals several traps and offers very many contradictions. It is all much less simple than we think. Today we can show that Laplace, when he promoted the idea of a deterministic universe, was very wrong.

Even simply imagining that, at a given moment and with infinite precision, we can know the conditions of every single particle that makes up our universe, would violate a whole series of principles of physics.

To simultaneously take a measurement of systems that are separated by tens of billions of light years means violating relativity; the only way of synchronizing activities so far apart in time and space would be by transmitting signals at unlimited velocity.

And then let's imagine, just for a moment, the complexity and dimensions of the apparatus needed to investigate the state of every single material component in the universe. A not insignificant fraction of the matter that makes up the universe would have to be organized into a measuring device. And who can guarantee that this transformation wouldn't have an impact on the outcome of the measurement itself?

We should then have to design and build an auxiliary measuring device, which monitors the material state of each single component of the main measuring device, as well as the scientists who are analysing the data and finally a third one which monitors the second and so on, to the point of creating a kind of immense fractal structure.

But even if it were possible to collapse, so to speak, all the wave functions at the same moment and immediately concentrate the information gathered into a massively powerful central data processor, the uncertainty principle would prevent us from knowing exactly, with infinite precision, the state of every material particle. Otherwise, we would have to violate the very fundamentals of quantum mechanics.

Even if, by some strange miracle, we were able to access the *instantaneous* quantum state of all the elementary particles in the universe, measured with infinite precision, this knowledge would not be of much help. In the subsequent behaviour of the particles subjected to measurement, the random mechanisms, which characterize the evolution of matter on a microscopic scale, would relentlessly recur. Ignoring, for a moment, that we have spied on them and forced them to freeze into a definitive state, they would very quickly continue to freely occupy all permitted states, without obeying any particular pre-established sequence. Nobody could predict precisely, moment by moment, the evolution over time of every single particle in the universe. We would have to be happy, once again, with average values and probability distributions, which would make the hypotheses that sparked this enormous effort futile.

Not to mention what happens in those zones that are inaccessible to us. It may be that entire regions of the universe, for example those hidden behind the event horizon of black holes, obey laws of physics that are unknown to us. It might also be the case that none of the logical or mathematical instruments, which we have developed to describe those quiet corners where conditions of equilibrium and stability are in force, would be of any use in understanding the laws that govern highly disturbed regions. Nobody can exclude the possibility that other inviolable principles exist, belonging to a physics which for now we do not know, and which is concealed in the regions characterized by the most turbulent and catastrophic phenomena in the cosmos.

And so, as modern physics celebrates the triumph of the materialist approach, it is, in fact, in the process of producing a vision of matter that is very different from the reassuring image usually attributed to it.

In certain respects, matter has been a marvellous illusion for human beings. Setting it against the world of ideas, which can't be seen or touched, has always been a kind of comforting refuge. The traditional materialist saw in it a stable, enduring component, something eternal and imperturbable, which forms a stable basis for everything. Matter, too, is subject to transformations, but the idea that atoms or particles, indivisible and indestructible, might explain the composition and behaviour of all material structures, from the tiniest to the most gigantic, has reassured humans for thousands of years.

The awareness that we were made from the same substance as the celestial bodies which light up the sky has consoled us and protected us from anxiety. Even the idea of death became bearable, when seen as a return to mother-earth. Our bodies may decompose, our identities vanish, but the material components will survive, returning to the womb of the great mother: indestructible, eternal matter.

The illusion that the fundamental components of matter are something solid and enduring has been seriously questioned by modern science. Not only can we not see, let alone touch, the elemental constituents, but they follow rules that are so different from those that govern the macroscopic world as to confute our every attempt to grasp them, even merely conceptually. They are ubiquitous waves which can be found anywhere and at the same time particles located at a very specific point; they are changeable material states which oscillate continuously between apparently different identities, but clandestinely united by deep symmetries, particles connected to other particles by weird relationships, sometimes one so far from the other as to seem unaware of the space that separates them, material states which appear and disappear from the vacuum at an infernal rhythm or which exchange interactions with everything that surrounds them, including the most distant material forms.

And if all this weren't enough, many of their properties, mass for example, depend on delicate mechanisms which could break apart from one moment to the next, producing catastrophes whose proportions are difficult to imagine.

Finally, everything, yes everything, is nothing more than a form of vacuum. What we had hoped might be a solid material basis, something to set against the evanescence of the idealistic approach, turns out to be just as impalpable, so fleeting as to risk being confused with a philosophical concept.

Our material world is made of vacuum. The conclusion that science has reached seems a bit of a joke. While trying to find something solid and material on which to base an explanation of everything, we've found ourselves removing any reassuring substance from the entire material world, and not only our world, but also the world of the most massive cosmic structures.

And what if particles were hiding unspeakable secrets from us?

Everything becomes even more complicated if we consider the world of particles from close up. There are too many things that don't add up. The suspicion that quarks and leptons might conceal unspeakable secrets has been circulating among us scientists for some time. Several fundamental questions have already been asked; for example, regarding the subdivision into fermions and bosons between particles which constitute matter and those which transmit forces. But there are many others, the answer to which might unleash big surprises.

Why three families of quarks and leptons and not four or five? Are the particles of the Standard Model truly elementary or do they have a concealed sub-structure? We don't know what gives mass to neutrinos, given that they are too light for us to hypothesize an interaction with the Higgs boson. What other mechanism came into operation and when?

And what can be said of the interactions, starting with gravity; why is it so weak? How does it operate over very small distances? We do not know what form it reduces matter into when it compresses it in a black hole. And, moreover, are the fundamental interactions which we have seen in action the only forces in the universe or are there other hidden ones?

As far as the fundamental constants of nature are concerned, such as the speed of light and Planck's constant, we don't know for sure where they come from. Is there a theory which allows us to extrapolate them? Are they truly constant or do they change over time? Does space-time have a microscopic structure? Do grains of space or time exist to make up its fine fabric? Are there only three spatial dimensions, or are there other hidden ones?

The list of unanswered questions could be much longer, but the one thing that is certain is that it will be the role of new experimental data to resolve the majority of such questions.

Sooner or later, it will happen that a young scientist studying the data alongside a group of their contemporaries, in a particle accelerator or in an astrophysics laboratory, will discover something wholly unexpected. That day, maybe, there will be an answer to one of these questions and that event could, once again, radically change our perspective on matter and the universe.

The beauty of our work is that we are sure that sooner or later this will happen; it's just that we don't know when. It might happen tomorrow, or maybe we'll have to wait fifty years for a new generation of brilliant young minds to grapple with the challenges of modern scientific research. It is to them that this book is dedicated.

Epilogue

Biella, 23 November 2021

I finally made it to Biella, but the car journey was somewhat trying. I left early in the morning because the meeting was scheduled for eleven o'clock. I am here through a combination of chance circumstances.

Everything began last summer, in August, when I was invited to speak about science at Arte Sella, an immense exhibition of contemporary art, an outdoor museum deep in a wild valley, near Borgo Valsugana in the province of Trento, in Northern Italy. That's where I got to know Emanuele Montibeller, the creator of Arte Sella, who accompanied me, after the lecture, on the long walk which connects the great works of art immersed in nature. And he was the person who organized this meeting.

Biella, city of wool, is proud of its thousand-year tradition. It is the textile capital, because still today, just under half of all the precious fabric manufactured in the world is produced here. This ancient vocation can be observed in the dozens of nineteenth-century woollen mills still present in this small town today. Some have become modern factories; others have been abandoned.

In one of these, the Fondazione Pistoletto set up *Cittadellarte*, a true citadel conceived as a place to defend art and plan the future of the community. And that is where we are going.

When Emanuele called me, some weeks ago, and told me that Michelangelo Pistoletto would enjoy having a chat with me, I couldn't believe it. I immediately accepted without a moment's hesitation.

And it wasn't just the unbounded admiration I feel for him; Pistoletto is one of the greatest living artists. Known throughout the world as the most authentic protagonist of Arte Povera, he has produced works which are exhibited in the most important museums and galleries. His *Venus of the Rags*, his mirror paintings, his *Terzo Paradiso* are among his best-known works and have made their mark on the collective imagination of the whole of humanity. But in the enthusiasm with which I greeted the proposal there was also something else, something very personal and profound. As if I had to meet Pistoletto to close the circle.

I had confirmation of this when, before tackling the subject that was closest to his heart, Pistoletto took me to see his very earliest works, which were quite different from the ones that had made him famous.

Pistoletto is amazed when I ask him questions about the tiniest details of the techniques that he was using then; I read the surprise in his eyes at my use of technical terms or because I want to touch the surface of the works and then observe them from close up to memorize all the details. When he asks where my curiosity, as well as a certain familiarity with spatulas, scrapers, acrylic bases, polyurethane foams and so on, comes from, I tell him about my father, Giuliano, the son of Guido the tailor.

My father was a railwayman by profession, but so passionate about art as to syphon off resources from the meagre family budget – this is the 1950s – to take drawing and painting lessons from the La Spezia masters of the Corrente group. The young stationmaster with a large family behind him develops a burning passion, sets up the artists' union, joins the Informale movement. From him I learned how to recognize the hallmarks of a given painter and understand their technique. His method of teaching was incredibly efficient. If, as a teenager, I told him I'd seen a canvas somewhere by Morlotti or Schifano which I'd liked, he took me into his little studio and made me experience with my own hands the secrets of their technique; in a few hours I watched two small pictures coming to life under my very nose, which even the most attentive critics would have struggled to attribute. Every effort he made to teach me the techniques which he had so skilfully mastered proved to be futile; and yet he passed on to me that love of art and beauty which has never left me.

When I tell Pistoletto all this, his eyes light up. 'My father was a painter too. I inherited my passion from him.' At that moment, I realize what has brought me here: to close the circle which the sudden death of my father, in 2011, had left open. The unexpected event had suddenly interrupted our ongoing discussion of art, which often started with the little pictures which he continued to paint, even though he was over eighty. In Michelangelo Pistoletto I found the same burning passion; it was a little bit like finding my father again.

Pistoletto is particularly interested in what contemporary science has to say about the origins of the universe. 'We'll never understand who we really are unless we understand the origin of everything.' This strong point of contact between the artist and the scientist is where the discussion starts: a discussion which I embark on, initially, with caution. I'm afraid of making statements which might in some way wound the artist's sensitivity or misconstrue the meaning of his research.

And so, we approach the heart of the problem slowly; but then I realize that, in fact, Pistoletto wants me to speak with extreme frankness, without compromise, about what modern science has to say about the origin of matter. That's why he called me, and I would be betraying his expectations if I overlooked any detail in his work which was in conflict with the most recent knowledge of contemporary physics. And so, I plunge right in.

Our lengthy discussions take place in the Universario, the part of the museum which is conceived as a kind of centre for research. The sign of the Third Paradise dominates the large whiteboard in the entrance. The word 'infinity' occupies the two outer circles, to express the connection between microcosm and macrocosm, while in the central circle 'space-time' is written to indicate the universe. I point out that an important element is missing here, because there are two material ingredients of the universe: space-time and mass-energy and both need to be referenced, both are necessary to an understanding of its birth. And we talk for hours about how space-time and mass-energy can emerge spontaneously from the void and produce a universe full of wonders.

Several months after our first encounter, we met again in Biella, and Pistoletto took me straight to the Universario. I noticed immediately that something had changed; in the formula of creation, the universe now developed through the combination of space-time and mass-energy. I couldn't not be moved to see everything that we know today about the birth of our universe depicted with the power of the imprint of a great artist. It affected me deeply to note that our chats had pushed one of the greatest artists of our period to modify his work. Pistoletto did it with humility and I was touched by that.

At that powerfully emotional moment, I thought how happy my father, Giuliano, the tailor's son, would have been if he'd been able to be there with us.

Index